Thomas Nichol

Diseases of the Nares, Larynx and Trachea in Childhood

Thomas Nichol

Diseases of the Nares, Larynx and Trachea in Childhood

ISBN/EAN: 9783337368326

Printed in Europe, USA, Canada, Australia, Japan

Cover: Foto ©berggeist007 / pixelio.de

More available books at **www.hansebooks.com**

DISEASES

OF THE

NARES, LARYNX

AND

TRACHEA

IN

CHILDHOOD.

BY

THOMAS NICHOL, M. D., LL. D., S. C. L.,

Member of the Colleges of Physicians and Surgeons of Ontario and Quebec; Member of the American Institute of Homœopathy, and Corresponding Member of the Homœopathic Medical Society of Pennsylvania.

NEW YORK:
A. L. CHATTERTON PUBLISHING COMPANY.
1885.

INSCRIBED TO

ALEXANDER THOMPSON BULL, M. D.,

OF BUFFALO,

Whom the author is proud to claim as his preceptor in " the art almost Divine."

PREFACE.

The essays composing this volume are the fruits of thirty years of study and experience. Some of them appeared in the *American Observer* many years ago, and now, encouraged by professional friends, they appear in book form, revised and enlarged. The pathology of each morbid state has been dwelt on at great length, simply because a correct understanding of the natural history of disease is indispensable to the scientific physician of any school. The homœopathic treatment is fuller and more minute than in any similar work, and the author's extensive experience enables him to speak with some little authority on this point. Should this volume be favorably received, it will be followed by another on the Diseases of the Bronchi and Lungs.

140 Mansfield Street, Montreal, March, 1885.

Table of Contents.

CHAPTER I.

ACUTE COYZA, ... 25
  Definition, 25 ; etiology, 26 ; symptomatology, 27 ; prognosis, 29 ; therapeutics, 29 ; general treatment, 37 ; aphorisms, 38.

CHAPTER II.

PURULENT CORYZA, ... 39
  Definition, 39 ; etiology, 41 ; symptomatology, 42 ; progress, 44 ; thermometry, 44 ; pathological anatomy, 44 ; diagnosis, 45 ; prognosis, 46 ; therapeutics, 47 ; aphorisms, 50.

CHAPTER III.

CHRONIC CORYZA, ... 51
  Definition, 51 ; varieties, 51 ; nature, 51 ; symptomatology, 52 ; progress, 53 ; thermometry, 53 ; diagnosis, 53 ; prognosis, 54 ; general treatment, 54 ; therapeutics, 54 ; aphorisms, 63.

CHAPTER IV.

SPASM OF THE GLOTTIS, ... 64
  Definition, 64 ; nature, 65 ; etiology, 67 ; symptomatology, 77 ; pathological anatomy, 82 ; diagnosis, 84 ; prognosis, 85 ; mode of death, 87 ; therapeutics, 87 ; general treatment, 110 ; chloroform, 111 ; tracheotomy, 111 ; aphorisms, 114.

CHAPTER V.

ACUTE CATARRHAL LARYNGITIS, ... 116
  Definition, 116 ; varieties, 117 ; history, 117 ; etiology, 118 ; symptomatology, 118 ; thermometry, 121 ; pathological anatomy, 122 ; diagnosis, 123 ; prognosis, 124 ; general management, 125 ; prevention, 125 ; therapeutics, 127 ; aphorisms, 131.

CHAPTER VI.

ACUTE ŒDEMATOUS LARYNGITIS, ... 132
  Definition, 132 ; etiology, 133 ; varieties, 133 ; history, 134 ; symptomatology, 136 ; progress, 137 ; duration, 137 ; scald throat, 138 ; pathological anatomy, 138 ; diagnosis, 140 ; prognosis, 142 ; therapeutics, 143 ; scarification, 149 ; tracheotomy, 150 ; aphorisms, 152.

CHAPTER VII.

SPASMODIC CROUP, ... 154
    Definition, 155 ; frequency, 157 ; etiology, 157 ; symptomatology, 159 ; auscultation, 162 ; diagnosis, 163 ; prognosis, 163 ; therapeutics, 163 ; general management, 168 ; prevention, 169 ; aphorisms, 170.

CHAPTER VIII.

PSEUDO-MEMBRANOUS CROUP, ... 171
    Definition, 173 ; history, 174 ; etiology, 177 ; recurrence, 181 ; heredity, 182 ; contagion, 185 ; symptomatology, 190 ; mechanism, 196 ; progress, 199 ; thermometry, 200 ; auscultation, 202 ; laryngoscopy, 203 ; essential nature, 204 ; pathological anatomy, 205 ; diagnosis, 210 ; prognosis, 212 ; general management, 214 ; tracheotomy, 215 ; therapeutics, 218 ; aphorisms, 233.

CHAPTER IX.

DIPHTHERITIC CROUP, ... 235
    Definition, 235 ; history, 236 ; symptomatology, 244 ; progress, 246 ; pathological anatomy, 249 ; identity or non-identity with pseudo-membranous croup, 252 ; prognosis, 266 ; tracheotomy, 268 ; general management, 274 ; therapeutics, 285 ; aphorisms, 286.

CHAPTER X.

SCARLATINAL CROUP, ... 286
    Definition, 288 ; symptomatology, 289 ; pathological anatomy, 290 ; diagnosis, 291 ; prognosis, 292 ; therapeutics, 292 ; prevention, 292 ; aphorisms, 293.

CHAPTER XI.

TRACHEITIS, ... 294
    Definition, 295 ; etiology, 296 ; thermometry, 299 ; pathological anatomy, 299 ; diagnosis, 300 ; prognosis, 300 ; general management, 300 ; therapeutics, 301 ; aphorisms, 302.

CHAPTER I.

ACUTE CORYZA.

Coryza is one of the most annoying of the affections of infancy and childhood, and, though in general mild, it is not destitute of danger to very young infants. It may be defined to be a catarrhal inflammation of the Schneiderian membrane —the mucous membrane lining the nasal cavity—for " the nose is the classical seat of catarrh." All the ancient physicians, including the Sage of Cos himself, believed that the secretion of the nose flowed down from the brain, a doctrine that was only exploded in 1660, when Schneider, of Wittenberg, whose name is still given to the mucous membrane of the nose, showed its erroneous nature. The phenomena of Coryza are of great interest to the physician from the fact that the diseased parts may be seen, and the changes which take place in the Schneiderian membrane are no doubt strictly analogous to those which take place in the mucous membrane lining the larynx and bronchial tubes.

Generally speaking, the mucous membrane of the respiratory organs is extremely sensitive in childhood, but it is a curious fact to which the attention of the profession was first called by Professor Jörg, of Leipsic, that this extreme sensibility does not exist during the first seven or eight weeks of life. "The exposure of an infant two or three weeks old to a low temperature or to a vitiated air will be followed by disturbance of the function of the liver, and the occurrence of jaundice; or, perhaps, the muscular power may be so far depressed as to render the child incapable of taking a full inspiration, so that its lungs collapse, and it dies from disorder of the respiratory organs, but without the cough and bronchitic symptoms, which would not fail, if it were a little older, to announce the irritation of the mucous membrane of the air tube. Why this is so I do not know, but I suppose it to be the result of the generally feeble vitality which renders the lining of the bronchia less susceptible; just as that of the intestine also seems to be at the same period, since, while constipation is frequent, diarrhœa is comparatively rare during the first two months of life." (West.) But it would appear that the nasal mucous membrane does not possess this happy insensibility, and during the first two months coryza is frequently annoying and at times dangerous.

Coryza generally results from the action of cold, from damp air of a low temperature, exposure to the weather, especially at the change of seasons, and especially is it caused by insufficient dress. Many people dress their little ones as if they possessed adamantine constitutions, and it is not uncommon to see a strong man warmly muffled up, walking with a shivering urchin very insufficiently clad. In infants it is sometimes caused from the extremities being chilled by the urine if the changing of the napkin is neglected. In some instances, infants are attacked with coryza so soon after birth that it would almost seem that they had been born with it. These cases are no doubt caused by the sudden transition from the warm temperature of intra uterine life to the trying and changeable circumstances of ordinary existence.

Coryza may also be produced by exposure to the heat of a strong fire or of the sun, and a "summer cold" is proverbially difficult of cure.

Acute coryza is a prominent symptom of the first stages of several infectious diseases, particularly of measles. As a general rule, acute coryza extends downwards, though sometimes the reverse holds good, and it may follow pharyngitis or laryngitis, and it is an occasional termination of bronchitis. Acute coryza is sporadic and epidemic but never contagious, though many hold that it may be communicated by using the same handkerchief, thus bringing the nostrils into direct contact with the morbid secretion. But so far, no one has yet succeeded in demonstrating the contagiousness of acute coryza by experiment, for Friedrich uniformly failed when he inoculated his own Schneiderian membrane with the secretion of people suffering from all stages of the disease. Children of all ages are liable to this affection, but it is most frequent in nurslings and children during the first years of life. As the infant grows older the liability to coryza increases, and the larynx and bronchial tubes are more likely to be involved. Female infants are more liable to the disease than males, from the fact that they are generally more feeble, and the foolish pride of dress causes them to be less efficiently clad.

The first symptom of acute coryza is sneezing, with dryness of the nostrils, causing a kind of snuffling respiration from which its common name of "snuffles" is derived, but, as Dr. Churchill points out, not every young infant which sneezes often has taken cold, for "the impression of light upon the branches of the fifth pair of nerves distributed to the eyes naturally gives rise to sneezing." Accompanying the sneezing is a fullness and swelling with tickling and itching of the nostrils. In a short time there is a copious secretion of watery fluid, colorless and transparent and of a saltish taste, which flows in a stream from the nostrils and which sometimes causes excoriation of the upper lip and the sides of the nose. This irritating quality seems to be due to ammonia, and in spite of the salt taste very little sodium chloride is present. As the disease advances the secretion becomes purulent or muco-purulent. The Schneiderian membrane is now vascular, tender and irritable, and the sense of smell is almost wholly suppressed. The child breathes by its mouth, and respiration is attended by a snuffling or rather snoring noise, which is almost pathognomonic. When this swelling of the mucous membrane is very considable the child is almost unable to suck, for the nostrils being closed, it is impossible to use the mouth for suction and respiration at one and the same time; of this any one may convince himself by trying to suck while compressing the nostrils tightly. The little one attemps to nurse, but in a few seconds the face darkens and it desists, crying and lamenting. Impelled by hunger, it again attempts to nurse, and after repeated efforts, it finally becomes exhausted with hunger, fatigue and suffering. The difficulty of swallowing is greatly increased if the catarrhal irritation should extend to the fauces.

The catarrhal irritation extends along the lachrymal passage causing the eyes to become red and watery, and later the disease extends to the frontal sinuses, causing a dull aching pain in the forehead which the child tries to alleviate by rubbing the forehead with its hand or boring the head into the pillow. As might be expected, there is a good deal of fever, the pulse is quickened, the skin is hot and some thirst is present. The senses of taste and hearing are less acute than in health, though they rarely suffer to the same extent as the sense of smell.

Such an attack is at its height in three or four days, after which it declines. The difficulty of breathing ceases, the discharge becomes thicker, and the danger now is that, if imperfectly cured, the disease may become chronic. For some time after recovery the patient is liable to relapses on very slight exposure to exciting causes.

The nasal mucous membrane is of a uniform red color, for the capillaries are surcharged with blood—at times it is reddened only in points. Soon all the subjacent tissues are infiltrated, and the mucous membrane becomes soft and swollen, and a slight swelling is sufficient to fill the nasal cavity, which is very small in young children.

Coryza is a comparatively trifling disease in children of four or five years of age, but it is a serious disorder in nursing infants. Dr. J. T. Meigs considers it "a serious and even dangerous disease," while Bouchut styles it "a very dangerous disease in children at the breast.' Dr. George B. Wood points out that on attempting to take the breast, children sometimes become black in the face from suspended respiration, and that they are said occasionally to be thrown into convulsions by the same cause. Both Wood and West admit that fatal cases occur from the difficulty of breathing and sucking. Fraenkel remarks that a fatal termination is extremely rare, and only takes place in nurslings owing to the disturbances of respiration and nutrition incident to closure of the nose. "While acute nasal catarrh is a complaint as common as it is harmless, it sometimes proves dangerous to infants at the breast, because the obstruction of their nasal passages, which are at all times narrow, makes it difficult for them to suck. If we do not feed with a spoon in such cases, life itself may be endangered in ill nourished or feeble children." (Felix von Niemeyer.)

Coryza responds readily to homeœpathic treatment, especially when it is aided by rational adjutants.

Aconite is, according to Doctor Hayward, the best remedy for the incipient stage of coryza. As a general rule, it is indicated by the symptoms during the first twenty-four hours, and, if promptly and persistently given, it often obviates the necessity for any other remedy. The indications are creeping chills, following exposure to dry, cold air or to a draught, and these chills are followed by burning heat, especially on the head and face; spasmodic sneezing and discharge of a thin, watery fluid from the nostrils, with great thirst, especially towards evening. A short, dry cough from irritation of the larynx is often present and profuse lachrymation is an almost invariable accompaniment.

The child is fearful and afraid during the day, and at night the sleep is restless and dream-haunted; the patient feels better in a cool room. A very small powder of the sixth or eighth decimal trituration should be given, dry on the tongue, every hour or every two hours. Hering does not recommend aconite for the primary disease, but "when the *catarrh has been suppressed* and headache is the result, give aconite."

Camphor is another useful remedy in the incipient stage ; indeed it is of little or no use unless it is given as soon as the morbid state appears. Dr. Hughes considers it more generally useful than Aconite. "A few doses of it rapidly dissipate that chilly feeling which with most persons is the precursor of a cold in the head." Camphor, then, is indicated by the chilly or cold stage when the malady is still in its incipiency ; shivering or coldness of the skin which at the same time is dry, and along with this there is heaviness, weariness and general malaise. Camphor is too much neglected, and this neglect probably arises from the fact that few practitioners carry the remedy in their pocket-cases. Place one drop of the Rubini tincture on a small piece of pure white sugar and give one-tenth of this every twenty or thirty minutes. I have seen decided good follow the repeated olfaction of the Rubini tincture. If the morbid state progresses, Camphor should be discontinued and some one of the under-mentioned remedies given.

Nux vomica has been much recommended for coryza, though Dr. Hempel remarks, "we have never been so fortunate as to effect anything great with Nux in a catarrh affection of any kind." On the other hand, Dr. Hughes says, "for the stuffy cold I think (herein again coinciding with Jahr) that Nux vomica is specific ;" and Dr. Ruddock agrees with the recommendation. Jousset recommends this remedy in the incipient, dry stage of fluent coryza, and says that by giving a dose of the third dilution every hour he has often arrested the malady by the end of its first day. Here I must range myself on the side of my old friend, time-honour'd Hempel, for I have rarely seen much good from Nux vomica in catarrhal affections. Hering recommends Nux for the same symptoms as Arsenicum when the latter causes no improvement in twelve hours, or when the catarrh is fluid during the day and dry at night. Nux vomica is usually given during the first stage when there is dryness or obstruction of the nose, with heaviness in the forehead and impatient mood; the mouth is dry and parched, without much thirst; tightness of the chest; constipation. Chills and heat alternate in the evening, and great heat of the face and head is present. Nux vomica has stoppage of the nose, especially out doors but fluent indoors, while the Pulsatilla coryza is fluent outdoors and stoppep indoors. This remedy acts best in small doses of the 30th dilution, given in the evening.

Mercurius is frequently given after Aconite, and is perhaps the most frequently indicated remedy in coryza. Hughes styles it "the established remedy," though personally he has a preference for Euphrasia, and Bæhr says that it is "a distinguished remedy which will scarcely be surpassed by any other." The symptoms are frequent sneezing, with soreness and redness of the nose, and constant watery discharge which gradually becomes purulent. The smell from the nostrils is often offensive, and the lips are swollen and excoriated. The eyelids are irritated with constant shedding of tears, and this irritation may extend to the air-passages, causing cough with mucous râles. There is alternate heat and shivering, the heat predominating over the chills, with profuse perspiration which affords no relief. The patient feels uncomfortable in a warm room, yet cannot bear the cold. Hering thinks that Mercurius is especially serviceable for children, and Ruddock says that it is often useful in alternation with Nux vomica, a recommendation in which I cannot concur for the simple reason that we have no proving of the two remedies *in alternation*, and if we had, the results would be worthless. Concerning the preparation of Mercurius to be administered, Teste lays down "that corrosive sublimate is indicated in an immense majority of the cases which have been considered until now as belonging to the sphere of soluble mercury, provided that, with a few exceptions, corrosive sublimate is given exclusively in the diseases of males, and soluble mercury in the diseases of females." Experience has amply confirmed this statement, and yet Merc. sol. acts well with children of either sex. Merc. corr. deserves the preference in coryza when the sneezing is excessive, and Mercury sol. when there is dull headache with great accumulation of mucus in the posterior nares. Mercurius acts well in all preparations, but I prefer small doses of the twelfth decimal trituration, dry on the tongue.

Hepar is of great service when the air passages are chiefly affected, when the cough is loose and croupy, with rattling in the chest, pain in the upper part of the windpipe while coughing, with hoarseness. The nose is often red and swollen, with scabby formations in the nostrils and loss of smell. Hepar is especially indicated when the catarrh is renewed by every breath of wind, or when it affects only one nostril and the headache is increased by every movement. This remedy is useful in most cases of ordinary catarrh after partial relief from Mercurius. Hering advises it when the symptoms have been better and became worse again, and Hayward uses it "to bring up the tone of the parts to its natural degree." I have had the best results from the 12th decimal trituration.

Arsenicum is of great service when the nostrils are stuffed up, with copious discharge of thin watery mucus, burning of the nose both externally and internally, with soreness of the adjacent parts. The nose is often swollen and there is frequent sneezing. The discharge is burning and corrosive, excoriating the upper lip and neighboring parts, and I have often verified the indication given by Jahr: "excellent if the nose is obstructed in spite of the copious discharge." There is foul smell from the nose and occasionally nosebleed is present. The patient is cold and chilly and the chills are intermixed with flushes of heat; general debility and prostration are almost invariably present. Hering remarks that this remedy is indicated "when there is not much fever, heat or thirst;" but I think, with Bæhr, that it is indicated "when the constitutional symptoms are very prominent and intense." The sufferings are relieved by warmth and exercise, and exposure does not aggravate the disease. The patient is thirsty, but drinks little at a time. The remedy is especially appropriate when the child has taken cold after a bath. As to dose, Bayes thinks that from the 3d to 30th will prove very serviceable, but after a long experience I find that I have had better results from the orthodox Hahnemannian 30th than from any other.

Chamomilla, though greatly neglected, is really one of the leading remedies for acute coryza in infants and young children, and Laurie remarks that "in the treatment of children this medicine is generally preferable to Nux vomica in arresting the attacks." Chamomilla is indicated when the coryza has arisen from suppressed perspiration, when a good deal of fever is present, one cheek being red and the other pale, chilliness and thirst are present, the temper is fretful and irritable, and the child wants to be carried all the time. The nostrils are often ulcerated and the lips chapped, and the discharge from the nose is copious and acrid. A hoarse cough is frequently present with rattling of mucus in the bronchial tubes, and this cough is worse at night, even during sleep. Chamomilla is doubly indicated if catarrh should make its appearance during dentition. Twelve years ago I wrote that Chamomilla should never be given lower than the twelfth dilution; I am now of the opinion that I should have written 'thirtieth' instead of 'twelfth.'

Belladonna is indicated by a dry, barking, spasmodic cough, coming on in paroxysms, apparently caused by titillation in the air passages, aggravated at night. Pain and heat in the head, eyes and nose are present, a throbbing, bursting headache, with flushed face and glistening eyes. The tonsils are swollen and red, with difficulty of swallowing and sensation of constriction in the throat. The breathing is short, anxious and hurried; the pain in the head causes the child to bore its head into the pillow or to rub it with the hands. The fever is quite high, with alternate chilliness and heat. The coryza is fluent, but the remedy is indicated not so much by the coryza as by the other symptoms. Behr recommends Belladonna if the tonsils are inflamed, and Jahr and Hering agree in recommending it if Hepar should prove insufficient. I have usually given from the 6th to the 12th dilutions, but have seen excellent results from the 30th.

Allium Cepa is indicated when catarrh is epidemic with much sneezing and running of the nose, which is inflamed and sore down to the upper lip; the nasal discharge is burning and excoriating. The eyes smart and burn, with profuse discharge of bland water; tingling and itching of the left nostril with violent sneezing. A laryngeal cough is also present, which increases towards evening. The catarrh commences mostly on the left side and moves to the right; it is worse at night and in a room, better in the open air and in the cold. Dr. Hering, in his lectures, was in the habit of insisting that Allium Cepa was the very closest simillimum to coryza, and he considered that it occupied a middle place between Aconite and Ipecacuenha. I have had the best results from the 12th centesimal dilution in repeated doses.

Euphrasia has been too much neglected in this disease, though it is more used at the present time than it was twenty years ago. "It acts upon the upper portion of the respiratory mucous membrane, *i. e.*, upon the conjunctival and nasal portions, only just reaching the larynx. It develops in this region a catarrhal inflammation, generally characterized by profuse secretion. Hence it takes a first place among the remedies for *fluent coryza* when this is a local affection, and not a symptom of general influenza, in which latter case Arsenic is preferable. The involvement of the conjunctiva in the catarrh is a special indication for Euphrasia, and sometimes the secretion from the eyes is acrid, while that from the nares is bland, the opposite condition obtaining with Arsenic." (Hughes). The Euphrasia coryza is violent and profuse, excessive discharge of white mucus from the nostrils, and this mucus, though generally bland, is sometimes acrid. The eyes are red and sore and the margins of the eyelids are occasionally ulcerated, with copious flow of tears. There is cough, but only during the daytime, and the entire disease is worse at night on lying down. Laurie recommends it to be given twelve hours after the last dose of Mercurius, if, after other symptoms having yielded, the flow of tears and cold in the head remain unmitigated, and I have often acted on this recommendation with excellent results. Hughes says that "small doses of the mother tincture, as recommended by Hahnemann himself, appears to answer all purposes excellently well." Ruddock recommends from the mother-tincture to the 3d decimal dilution. Bayes says "the disease is readily cured by Euphrasia 6 or 12." My own experience has been made with the 3d decimal dilution.

Pulsatilla is frequently suitable after Chamomilla, and, according to Bœhr, " it may afford more relief than any other remedy when infants at the breast are attacked with catarrh, which, even if it runs its ordinary course, becomes a source of distress because it prevents them from nursing." It is indicated by a flow of thick, yellowish, fetid mucus, swelling of the nose with ulceration of the nostrils, frequent sneezing and roughness of the voice. The child is chilly in the evening and has whining moods, with loss of smell and appetite ; feels better in the fresh air, worse in the warm room. Jahr remarks that " Pulsatilla is appropriate if, after the mucus begins to assume a thicker consistence, the nose is alternately stopped and running; this remedy is scarcely ever indicated as long as the discharge is watery, but is very often better adapted to the case than any other medicine if the nose continues to discharge for an undue length of time a thick yellow or green mucus, and likewise if the nose is only obstructed in the evening and in the room, and runs again in the open air." " When nasal catarrh has passed into its third stage of thick and bland discharge, and is inclined to linger, Pulsatilla is the medicine best calculated to hasten its departure, and may be relied on no less in chronic coryza of simple character and without constitutional taint. It will cure even when the flux is so profuse as to deserve the name of rhinorrhœa," and the present writer has succeeded in effecting cures in a number of cases in which the disease had affected the frontal sinuses, with very offensive discharge. Like Chamomilla, this remedy acts best in the much-ridiculed thirtieth dilution.

Sambucus is suitable for new-born infants; the nostrils are obstructed by a thick, tenacious mucus; the throat and mouth are dry, and the nostrils seem to be completely closed, yet no thirst is present. The child has sudden startings from sleep as if suffering. Sambucus is the only remedy mentioned by Hempel for this disease in children. Noack and Trinks recommend one drop of the mother-tincture or of the 1st, 2d or 3d dilution once or twice daily, and it is almost always given in low dilutions.

Dulcamara is the most appropriate remedy for children who are subject to severe coughs, or to sore throat whenever they are exposed to a damp atmosphere. The patient feels better when in motion and worse during rest, and the slightest exposure renews the obstruction of the nose. In such cases Dulcamara is preventive as well as curative, and it acts best in the lower dilutions.

Other remedies are Carbo veg. for fluent coryza with hoarseness and rawness of the chest; Arum Triph. for acrid fluent coryza excoriating the nostrils and adjacent parts; Cyclamen, the Pulsatilla of chronic diseases, according to Hering, for frequent sneezing with profuse discharge; Ammon. carb. for dry coryza with stoppage of the nose; Sang. Canad. for fluent coryza with cough and diarrhœa; Ipec. when there is difficulty in breathing, "give a couple of times" (Hering); Bryonia for hard cough with soreness of the chest, also for difficulty of breathing if Ipec. does not relieve (Hering).

Hippocrates recommended that the nose should be greased with the view of alleviating the difficulty of breathing, and this simple expedient is just as effective now as it was two thousand years ago. Where the nostrils are dry and obstructed, injections of glycerine and tepid water afford relief. "During the first day or two *steaming* the head and face will afford great relief, especially if a few drops of Aconitum be added to the water; and whilst giving the *mercurius* internally the same medicine may be used as spray, warm (5 grs. of the 1st trituration to 8 oz. of water); the patient should be kept to one room, and the air should be kept warm, 65°, and moist by having steam continually escaping into it." (Hayward.) "If the nasal obstruction is such that it entirely prevents respiration and suction, the physician should attempt the introduction of a small silver tube into each nostril; it should be flattened, and curved from before backwards following the course of the floor of the fossæ, and afterwards fixed under the nose with the neighboring tube. These two provisionary canulæ allow the passage of air, and prevent the child from dying at once, by giving the disease time to cure itself." (Bouchut.)

Dr. Charles D. Meigs directs a flannel cap to be put upon the child and worn for two or three days. The cap should be removed as soon as the coryza is relieved, as otherwise the child is apt to become so accustomed to it as to take fresh cold when it is removed. Dr. Ruddock thinks that infants should be taught to breathe through the nostrils, especially during sleep, but I fear that is "easier said than done."

Of all the auxiliary measures that have been recommended, I have found the "thirst cure" the most efficient. It was first suggested by Dr. C. J. B. Williams in the *Cyclopædia of Practical Medicine*, and is a most powerful means of cure. "It is the acrimony of this discharge (from the pituitary membrane) which reacts on the membrane and keeps up the inflammation and its accompanying disagreeable circumstances. On this circumstance depends the efficacy of a measure directly opposed to that just noticed, but to the success of which we can bear decided testimony—we mean *a total abstinence from liquids*. This method of cure operates by diminishing the mass of fluid in the body to such a degree that it will no longer supply the diseased secretion. The coryza begins to be dried up about twelve hours after leaving off liquids; from which time the flowing to the eyes becomes gelatinous, and between the thirtieth and thirty-sixth hour ceases altogether. The whole period of abstinence needs scarcely ever to exceed forty-eight hours." The thirst cure is, of course, not suitable for nursing children, but for those who are past that stage I know no better aid to the homœopathic remedies.

Dr. Constantine Hering in his lectures was much in the habit of dwelling upon the connection between the use of *salt* and *sugar* on the one hand and coryza and catarrhal affections on the other. "If a patient is subject to very frequent recurrence of catarrhs which are very difficult of cure, it will often be found that he eats too much salt. In this case he should be as moderate in the use of salt as possible, and smell now and then sweet spirits of nitre." And again, "Never suppress a cold either by cold or drugs, it is always a purifying process. Nobody takes cold who has not other impurities in his system. One is much more liable to catch cold after eating or drinking sharp, superfluous or indigestible things. Many children will not get rid of a cold as long as they indulge in too much sugar, syrup and other sweets."

Aphorisms.

1. Snuffling of the nose, with its attendant difficulty in breathing, at once calls our attention to coryza.

2. Coryza is not a dangerous disease, save in the case of feeble infants, for fatal cases undonbtedly occur from the difficulty of breathing and sucking.

3. Aconite and Camphor are the leading remedies for the early stages, and Mercurius and Arsenicum for the more

advanced, but Euphrasia, Chamomilla and Allium Cepa have hitherto been too much neglected.

4. The thirst cure is the most effective of the accessary means of cure, and, according to Constantine Hering, the immediate use of salt is one of the chief predisposing causes, while the immediate use of sugar hinders the cure.

CHAPTER II.

Purulent Coryza.

Purulent coryza, called by Underwood coryza maligna or morbid snuffles, is an inflammation of the mucous membrane of the nose in infants which, still being acute, differs from simple coryza in the much graver character of its symptoms. Its distinguishing feature is the presence of a purulent secretion accompanying an inflammation of a more or less malignant nature; Dr. Fraenkel, who has given us an excellent essay on this disease, thinks that it would be etymologically more correct to call it a pyorrhœa than a blennorrhœa. This morbid state rarely occurs alone, being often associated with angina, conjunctivitis or otitis, and in my own practice I have seen purulent coryza as a sequel of scarlet fever, diphtheria and measles. In these cases a pseudo-membranous exudation is present which is frequently not to be distinguished from a diphtheritic membrane. In illustration of this variety of coryza I quote the following instructive case, occurring as a complication of scarlatina, from Dr. Charles West: "In this instance, a little boy, six months old, was brought to me on the 25th of October, 1842. His health had been good until the 20th, when he became hoarse; on the 22d this hoarseness had much increased, and he became unable to suck, since which time he had continued to grow worse. When I saw him his skin was warm, face rather flushed, eyes watering, and a thick, ropy musus obstructed his nostrils. He cried with a suppressed but squeaking voice, and breathed with a peculiar wheezy noise, though air entered the chest unattended with any râle. The child was unable to suck, and even when he drank from a cup the fluid often returned through his nose. The inside of the mouth was very red, and the tonsils and soft palate were especially so. The mouth was full of an extremely tenacious mucus, which it was necessary from time to time to take out with the hand. A lotion was injected up the nostrils, composed of ℨj of alum to ℥ij of water, with great relief to the child, the secretion from the nares becoming more decidedly puriform, but less adhesive; and the child became able to suck a little. On the 28th, however, the child's powers seemed much depressed; it sucked eagerly, for the secretion from the nose had become almost watery, but it swallowed with much difficulty. A layer of false membrane of a yellowish-white color had now appeared on the soft palate and back of the hard palate, and on the tonsils. A lotion of three grains of nitrate of silver to an ounce of water was applied to the back of the throat, and a mixture of the extract of bark with ammonia was given every six hours. On the first of November the child was better, could both swallow and suck well, and the false membrane had entirely disappeared from the mouth; but the palate was still red, and presented some broad superficial patches of ulceration. The subsequent recovery was tardy, but the immediate danger was over, and no relapse occurred." No one can doubt but that at the present day this case would be pronounced diphtheritic by well-read and experienced practitioners of all schools, and probably Dr. West himself would now be of that opinion. Denman describes an epidemic of this disease under the name of *coryza maligna,* and he states that in connection with the coryza there was a general fulness of the throat and neck externally; that the tonsils were tumefied, and of a dark-red color, with ash-colored specks, and in some cases, with extensive ulcerations; and that some of the children swallowed with difficulty. Meigs and Pepper, in commenting on Denman's remarks, say, "there can therefore be little doubt but that in reality these were cases of nasal diphtheria," but most observers would omit the adjective nasal and simply call them diphtheritic.

Purulent coryza chiefly occurs on the Continent of Europe, especially in the Foundling Hospitals; less frequently is it found in Great Britain, while on this continent the disease is so seldom seen that many practitioners have never seen a case. On the whole, then, purulent coryza is a somewhat rare disease, so much so that Rilliet and Barthez, the leading French writers on the diseases of children, do not even mention it. It affects both sexes with like frequency, and is usually seen in the newly-born, and hardly ever in children over twelve months.

The causes of simple acute coryza, exposure to a damp or cold atmosphere or neglect in changing the infant's clothing, have but little influence in causing the much more serious purulent coryza, though unquestionably a certain proportion of cases arise from accidental aggravations of the milder disease, and I attended a case in which this result was produced by exposure of a child suffering from the simple form of coryza to the heat of a strong fire. But the most influential factor in the etiology of this disease is undoubtedly the infection of the mucous membrane of the child's nose with the secretions of the maternal vagina during birth, so that the disease is strictly analogous to *ophthalmia neonatorum*. This view of the chief cause of the disease is confirmed by the facts that it almost invariably appears during the first days of life; that it rarely attacks children whose mothers are not suffering from leucorrhœa or some similar malady: and, lastly, that it is most frequent in little ones who, from various causes, have been long detained in the maternal passages during birth. Then again, in the great majority of cases, all other causes of the disease, save infection by maternal secretions, can be excluded with almost absolute certainty. It is worthy of remark that ophthalmia neonatorum is much more frequent than purulent coryza caused by the maternal secretions, and this comparative immunity of the nasal passages probably arises from the movements of the eyelids favoring the entrance of the infecting matter in the one case, while in the other the ciliated epithelium of the nasal mucous membrane probably acts as a protector against the *materies morbi*.

As soon, then, as the child is born, or at least a very brief period after birth, a watery, bloody discharge from the nose, accompanied by sneezing, announces the onset of the disease. Sometimes stoppage of the nose precedes the discharge, but as a general rule the latter, which is the pathognomonic symptom of the disease, makes its appearance first. This discharge is usually yellow in color, odorless, at first glutinous, but soon it becomes thicker and purulent with a peculiar smell, which, however, differs from the fetid odor of chronic coryza. At times it resembles the "laudable pus" of the older surgeons, at times it has the color of prune juice from an admixture with blood, but, as a general rule, it is purulent, rarely mucous. Later it is thickish, rather solid in consistence, especially when the cause of the disease is infection from the maternal secretions. A thin, ichorous discharge, containing at a later date small granular particles—really the detritus of the pseudo-membrane—indicates the presence of false membrane, which, on examination by a strong light, is seen covering the nasal mucous membrane with a uniform yellowish-white coating. Soon the alæ nasi and adjoining parts are inflamed and swollen, and the red and shining skin is really the seat of an erysipelatous inflammation. The upper lip is red and swollen, and, at a later stage, it is excoriated by the secretions. I have never noticed "the curious purple streak on the margin of the eyelids" which Denman considered to be pathognomic; and in many cases there is a fulness and swelling about the throat and neck externally, resembling the well-known enlargement of diphtheria.

The swelling of the nasal mucous membrane is much greater than in the simple coryza, and, as a result, the breathing is difficult, nasal and snoring. Young infants breathe almost exclusively by the nostrils, and when these are plugged by the inspissated secretion of this disease, they seem to be quite unable to keep the mouth open in order to compensate for the closure. When the mouth is kept open it, together with the tongue and throat, becomes dry and stiff, and the infant makes such violent efforts to breathe as to conduce greatly to a fatal termination to the disease. When the nostrils are closed the child, unable to breathe and suck at the same time, refuses the breast, or only nurses at considerable intervals and with great difficulty. Cough is rarely present, except in those cases in which the disease has extended to the fauces; bleeding from the nose sometimes occurs in the pseudo-membranous form of the disease.

The general appearance of the child, from the very beginning, indicates a serious malady, quite different from even severe cases of simple acute coryza, for in addition to the intense inflammation of the entire nasal mucous membrane, there is great constitutional debility present in almost all cases. The violent attacks of dyspnœa are, of course, the result of the closure of the nostrils, and the restlessness, depression and emaciation are the expression of the constitutional disease. The skin becomes dry and harsh, and as the emaciation progresses it becomes wrinkled, and low fever and somnolence are frequently seen in advanced stages. I have never met with the disorder of the bowels, with thick, pasty stools of a green or blue color, of which Fleetwood Churchill speaks.

In favorable cases the inflammation with its accompanying discharge diminishes, the swelling of the nasal mucous membrane—which carries with it so much of danger—subsides, breathing becomes quite easy, and the child soon enters upon convalescence. Of course, as soon as the nasal passages permit free respiration the act of sucking becomes easy, and with this the debility and emaciation soon pass away. But all cases have not this favorable termination, for the little one may perish from inanition caused by pain, fatigue and insufficient nourishment, and the fatal result is ushered in by drowsiness which soon deepens into coma. Death, in these cases, is often caused by effusion on the brain, and, indeed, severe brain symptoms are frequently associated with purulent coryza.

Violent cases of this disease may prove fatal in three or four days, and milder cases may run on for a week or ten days before amendment takes place, but even then final recovery only comes after careful treatment of the destructive processes in cartilage and bone so apt to follow severe ulcerative inflammation. The duration of the disease depends greatly upon the age of the patient, for in very young infants the fatal termination is always nearer at hand and always more threatening than when the patient is at least a year old.

In severe cases of purulent coryza, the thermometer shows a temperature of 101° to 104°, and such high temperatures add very greatly to the danger. Still, in one malignant case which I attended in the year 1860, the temperature was quite low throughout, and a moderate temperature is no guarantee for a favorable termination, especially if the disease tends to assume the malignant form.

Dr. Denman, the celebrated accoucheur, met with an unusual number of cases of purulent coryza, encountering eight cases in eight months, of which six died. One of the bodies was opened by John Hunter and Sir Everard Home, who detected nothing save that the nasal mucous membrane was of a dark-red color, and its blood-vessels more turgid than usual. Later observers note that the mucous membrane is softened as well as thickened throughout the entire extent of the nasal fossæ, and that the membrane is thickly coated with pus or a thick, tenacious mucus. On removing this mucus exuded blood is seen, mostly in minute points, which were thrown out in the course of the disease. In other cases patches of the pseudo-membranous exudation are found scattered over the surface of the nasal mucous membrane, and this exudation is not necessarily diphtheritic. In other cases again the swollen mucous membrane, of a vivid red hue, is covered throughout its entire extent with a closely-adherent pseudo-membrane which extends over the entire interior of the nares, and these are the cases which it is almost impossible to distinguish from diphtheria. On removing the pseudo-membrane the subjacent mucous membrane is found to be softened and extremely ulcerated. Bouchut remarks that very commonly the false membranes are not situated in the interior of the nasal fossæ, but only at the orifice of the nostrils. Purulent coryza can hardly be confounded with simple acute coryza, for the violent inflammation with purulent discharge of the first-named is entirely different from the catarrhal inflammation with mucous discharge of the other. The differential diagnosis between purulent coryza and diphtheria is much more difficult, and in the advanced stages of the disease it mainly rests on the presence or absence of the characteristic diphtheritic blood-poisoning. If a false membrane is distinctly visible in the nasal passages, or if the nasal discharge is loaded with minute fragments of false membrane, and if this is followed by the well-known symptoms which mark the constitutional infection of diphtheria, there can be no room for doubt, and the disease is diphtheria beyond a doubt. But when a pseudo-membrane lines the nasal passages without constitutional symptoms following, it is likely that the disease is not diphtheria, for not all false membranes are diphtheritic, and most experienced practitioners have met with cases of pseudo-diphtheria which present a most wonderful resemblance to the genuine disease. Again, isolated patches are likely to be non-diphtheritic, while a continuous coating of false membrane is almost certainly diphtheritic. Abscess of the nose has some resemblance to purulent coryza, but abscess rarely appears on both sides of the nose, while purulent coryza, as a very general thing, affects both sides with like virulence. Then the course of the diseases differs much, for in nasal abscess the one-sided inflammation is, after a few days, relieved by a discharge of pus which brings welcome repose to the patient, while in purulent coryza the virulent inflammation is accompanied by purulent discharge almost from the beginning. Purulent coryza has been confounded with croup, though it is difficult to understand how any one could make the mistake, for the whistling inspiration and sudden dyspnœa which follow the closure of the nasal passages are wholly unlike croup, and the application of the stethoscope to the larynx, which should never be omitted in croup and, indeed, all laryngeal diseases, soon makes the case clear. Purulent coryza may possibly be confused with syphilitic coryza, but the history of each case must be carefully investigated, the entire course of the disease is different and the characteristic eruptions soon clear up the diagnosis. Later it will be confirmed by the changes in the shape of the central incisors of the upper jaw, first clearly pointed out by Jonathan Hutchinson.

Purulent coryza is always a serious disease, and the danger in each particular case depends much upon the degree of tumefaction of the nasal mucous membrane and upon the consistence of the secreted fluids, for upon these two factors of the disease depend the ability to breathe and to suck. Denman lost three-fourths of his cases, and Meigs and Pepper say that "the two cases of idiopathic membranous coryza in infants that came under our observation both proved fatal," while "the four cases in older children recovered without any difficulty." A good deal depends on the age of the patient. A feeble, newly-born babe offers little or no resistance to a severe attack of purulent coryza, but a stout little one, of say nine months, may get through even a severe attack. The pseudo-membranous form is more dangerous than the purulent, for, as has been already remarked, many cases of the pseudo-membranous variety are really diphtheritic, though perhaps an equal number are strictly analogous to the well-known pseudo-diphtheria.

The nasal secretions should be removed as they collect,

though this would be bad practice if a tightly-adherent pseudo-membrane is present. To soften the secretions a very small quantity of tepid water may be thrown into the nares by means of a small syringe, and the passages may then be cleansed with a camel's hair pencil. Fraenkel remarks that infants must not receive injections into the nose while lying down, as in this position the medicated fluids are very apt to pass through the pharynx into the opening of the larynx, producing severe spasm of the glottis. Indeed, the child should be kept as much as possible in an upright posture, and the little one's mouth should be kept open in severe cases. When the nostrils are completely blocked the child should not be put to the breast, but the maternal milk should be given from a spoon. It is in this disease rather than in simple acute coryza that the nasal tubes of Bouchut, mentioned in the preceding chapter, are useful. Tubes of soft rubber would be preferable to silver ones, though the latter would suit best if the nasal secretions are both thick and firm.

I am not aware of the existence of any essay on the homœopathic therapeutics of purulent coryza, so I will first state my own experience in the disease, and then briefly indicate the remedies which will cure every curable case, for some cases of this disease are incurable in that very nature. My experience includes three well-marked cases, and these were cured with Argentum nitricum, Nitric acid and Apis mellifica, each with a single remedy, unaided by any adjurant save attention to cleanliness.

The disease has, save in the well-marked pseudo-membranous form, an etiology which is identical with that of ophthalmia neonatorum. Now Argentum nitricum is the leading remedy for ophthalmia neonatorum, and Dr. Hughes writes as follows: "I myself have been so satisfied with even its internal effects in ophthalmia neonatorum that I have never had to resort to any external measures beyond those needed for cleanliness." The experience of our American oculists is quite confirmatory of its power over such purulent inflammations of the conjunctiva. Dr. Angell commends the remedy "in affections of the lining membrane of the lids, and of the lachrymal duct and sac, when there is an abundant discharge of pus;" and Drs. Allen and Norton write: "The greatest service that Argentum nitricum performs is in purulent ophthalmia. With large experience in both hospital and private practice, we have not lost a single eye from this disease, and every one has been treated with internal remedies, most of them with Argentum nitricum of a high potency, 30th or 200th. We have witnessed the most intense chemosis with strangulated vessels, most profuse purulent discharge, even the cornea beginning to get hazy and looking as though it would slough, subside rapidly under Argentum nitricum internally."

In my first case the child was two months old, weak and scrofulous; the mother had suffered for years from a very profuse leucorrhœa. Ophthalmia neonatorum was present as well as purulent coryza, so that there could be no doubt as to the etiology of the disease. The nose was red, swollen and painful, especially over the nasal bones. Several pimples studded the tip and neighboring parts, and these red and angry pimples speedily opened and became small ulcers. The discharge was thick, yellow and blood-streaked, and twice a small hemorrhage made its appearance. The child was gloomy and sad, and the entire state was worse after midnight and also in the morning. I gave Argentum nitricum, 12th centesimal dilution, and both ophthalmia and coryza were cured in a fortnight. No external applications were used, save abundance of tepid water.

In my second case there was a very strong suspicion, almost amounting to certainty, of a syphilitic taint. The child was four and a half months old, wan and withered, with pinched features and skin drawn tightly over forehead and cheek-bones. The mucous membrane of the nose was ulcerated, with a constant discharge of thin, bloody, fetid sanies, which corroded the upper lip. The nose was of a vivid red, and studded with small yellowish vesicles which broke and formed scabs. As the disease advanced the nasal discharges became thick and yellowish, but the streaks of blood disappeared. The fetid smell lingered to the last, and we had three somewhat profuse hemorrhages of dark blood, all in the night-time, without any special cause. The aversion to the open air was very marked, but cold weather agreed best with the child. I gave Nitric acid, 12th centesimal dilution, and the child improved at once, though it took three months to complete the cure. No other remedy was used.

My third case was cured with Apis mellifica, 6th decimal trituration. It was a well-marked specimen of the pseudo-membranous variety of the disease, the nares being coated with a false membrane which yet was not diphtheritic, for it lacked the fetor which is almost part and parcel of diphtheria, and no constitutional symptoms accompanied or followed the local disease. The nose was greatly swollen, red and œdematous, and so marked were these external symptoms that the relatives of the child at first thought that the disease was erysipelas. The mucous membrane swelled to such an extent that the nose was completely stopped up —even before the appearance of the false membrane. The inflammatory action was followed by the exudation of a tenacious, gluey mucus, which speedily became organized into a well-marked false membrane, on removing which the subjacent mucous membrane was seen to be still swollen and studded with minute bleeding points. The morbid action extended to the fauces and even threatened the larynx, but finally made a good recovery in twelve days.

Aconite would only be of value in purulent coryza if administered very promptly, almost before the morbid state had time to develop itself, as it were; if given afterwards it would cause the loss of valuable time. Belladonna is more frequently indicated than Aconite, corresponding as it does not merely to the symptoms of the malady, but to the pathological state of which the symptoms are the expression. Leading remedies are Mercurius solubilis, Hepar sulphuris, Arsenicum album, Calcarea carbonica, Pulsatilla, Sulphur,

Silicea, Aurum muriaticum, Lachesis and Kali bichromicum, the two last mentioned being especially effective in the pseudo-membranous form of the disease. For the special indications I must refer the reader to Chapters I and III of this volume.

APHORISMS.

1. Purulent coryza is a malignant inflammation of the nasal mucous membrane of infants characterized by a profuse purulent discharge and, at times, the formation of false membranes which yet are not diphtheritic.

2. Purulent coryza is chiefly caused by actual contact of the infant's nose with morbid secretions of the maternal passages during birth, and hence the disease is closely analogous to ophthalmia neonatorum.

3. Purulent coryza is comparatively a rare disease, and the mortality, even under the most enlightened treatment, is probably at least one-half of the whole number attacked.

4. The pseudo-membranous form of purulent coryza is distinguished from true diphtheria of the nasal passages by the presence of constitutional infection when the disease is diphtheria, and also by the fact that true diphtheria rarely attacks the nose alone.

5. The remedies which have proved successful in the writer's hands are Argentum nitricum, Nitric acid and Apis mellifica; other remedies are Sulphur, Mercurius solubilis, Arsenicum album, Aurum muriaticum, Lachesis and Kali bichromicum, the two last named especially in pseudo-membranous cases.

CHAPTER III.

CHRONIC CORYZA.

Chronic coryza of infants is comparatively rare, but as it is exceedingly difficult of cure, it is advisable to describe the disease and its treatment as fully as possible. This intractability arises from the constitutional taints which so often lie at the root of the local affection, which, in these cases, is merely a manifestation of a constitutional disease.

Chronic coryza may be defined to be the morbid state which follows a neglected or partially cured acute coryza. Ulceration may be present, but simple, chronic inflammation of the Schneiderian membrane is the most common pathological state. There are then several varieties of chronic coryza. The most common of these is the simple form dependent on chronic inflammation, and the constitutional state here is a low condition of health, with mal-nutrition and anæmia. Next in frequency we have the scrofulous variety, exceedingly intractable in its nature, but still quite amenable to homœopathic treatment; and the syphilitic variety, the most formidable of them all, but which, as Bouchut, long ago pointed out, is cured more easily than the others.

Chronic coryza is not nearly so common as acute coryza, and, if all cases of the last mentioned were carefully attended to, the chronic variety would become still more rare. So far as I have observed, it is not more frequent in children of one sex than of the other. Some fault of the general health, some obscure constitutional dyscrasia is almost invariably the predisposing cause of chronic coryza, and Fraenkel remarks " that acute rhinitis may pass into the subacute and chronic forms, and yet in the vast majority of cases this only takes place in *persons suffering under a dyscrasia.*" In not a few cases, however, children not suffering from any dyscrasia may have repeated attacks of acute coryza which finally terminates in the chronic form.

The principal symptoms of chronic coryza, as might be expected, are of a strictly local character. The respiration is nasal, and embarrassed even during the day, and at night the obstruction of the nostrils gives rise to snoring, or rather hissing sounds. The child's rest is disturbed by the necessity of making increased muscular effort to fill the chest with air, and as a result, the sleep is broken and restless. In aggravated cases the difficulty of breathing is so great that the blood becomes so thoroughly carbonized, that the sleep is heavy and restless. On examining the nasal passages the mucous membrane will be found to be thickened and injected. In the earlier stages it is more highly vascular than natural, and here and there slight excoriations are visible. As the malady advances the mucous membrane becomes pale, bloodless, and devoid of its natural velvet-like lustre. In many cases the amount of secretion is so much smaller than in acute coryza that they are spoken of as "dry catarrhs," while in others the secretion is purulent and very abundant. Whether scanty or copious, the secretion is so viscid as to form scabs and crusts, or even small lumps of inspissated mucus, and this combined with the thickened state of the mucous membrane causes a true stenosis of the nostrils. These crusts are moist and greenish in the earlier stages, and dry and blackish in the more advanced phases of the disease, and if they contain blood they are dark reddish in color and friable in texture. Very little causes the child's nose to bleed, and these frequent hemorrhages often cause the physician to be consulted. As the disease advances the voice changes and becomes markedly nasal. In chronic coryza we do not find the prickly itching in the nostrils, the sneezing or the frontal headache that are so prominent in acute coryza, though in some instances the child would seem to suffer from frontal headache, judging from the manner it rubs its forehead against the nurse's shoulder. The decomposition of the secretion gives rise to a more or less intense smell from the nostrils and even from the mouth, a peculiar odor given off with the expired air, and when this symptom is present, the disease is called *ozæna*.

As the disease advances, the general appearance of the patient gives evidence of greatly impaired health. The face is pale, the complexion is dusky, the features lose their lively expression, and all the movements of the child show languor and listlessness. The sleep is unrefreshing, the appetite is capricious and finally fails altogether, and nutrition becomes seriously impaired. The tongue is pale and flabby and more or less coated, and either constipation is present, or constipation alternates with diarrhœa.

Chronic coryza is one of the least self-limited of all diseases, running on indefinitely till cured. It is useless to look for a cure short of several months, for there is a strong predisposition to acute exacerbations of the original disease.

In the earlier stages of chronic coryza the thermometer shows little alteration of temperature, but as the disease advances a kind of mild, hectic fever is developed and the evening temperature rises one or two degrees. This mostly occurs with delicate children in whom there is possibly a suspicion of scrofula. During the acute exacerbations the temperature rises, as a matter of course.

The nostrils should be carefully examined by a full light, and, at the same time, the fauces should be examined with equal care. In infants, rhinoscopic examinations are almost impossible, and it is not always easy to get older children to submit to them, but they should be instituted whenever practicable.

The diagnosis is easy if proper care is used in the examination. The teeth should always be examined, as the foul odor may arise from dental caves, and collections of matter in the follicles of the tonsils may give rise to similar symptoms. The differential diagnosis between the several varieties of chronic coryza—ulcerative, syphilitic, and scrofulous—must be based upon the most careful and thorough investigation of the history of each case; for the syphilitic and scrofulous varieties are only to be distinguished from the form dependent on simple chronic inflammation *by their histories*. A seated pain in the cheek or forehead would indicate extension of the disease to the antrum of Highmore or to the frontal sinuses. Young children would make this known by rubbing the affected. parts against the pillow or against the nurse's shoulder.

Very few children die from chronic coryza, so that a favorable prognosis must be given as far as life is concerned, but it is a most obstinate disease, and, even under the most enlightened homœopathic treatment, it requires a number of months to effect a cure. A good deal depends upon the stage of the disease at which the patient comes under treatment ; for if the case is seen early the prognosis is more favorable than if the case has progressed to atrophy of the nasal mucous membrane.

In the management of the chronic coryza of infant's, clothing is even more important than diet. Flannel underclothing must be insisted on during the cold months of the year, and merino underclothing should be worn in summer. The underclothing, even in summer, should come high up on the neck, and both upper and lower limbs should be protected. for countless cases are. caused and finally goaded into incurability by the foolish custom of leaving the legs of tender infants almost naked. The diet should be plain, but nutritious, and all rich foods should be carefully avoided.

Sulphur is a leading, in fact an almost indispensible remedy in chronic coryza of children. It is indicated in weakly children of psoric constitution, for almost all the little ones helped by this great polychrest have suffered from eruptions on the skin, or from diarrhœa. In such patients the skin is unhealthy, and every little injury inclines to suppurate and to heal slowly. The nostrils are excoriated and ulcerated, with profuse discharge of thick, yellowish or greenish puriform mucus, and frequently the nose is obstructed by hard, dry scabs, with frequent bleeding from the nose. In almost every case the nasal discharges have an offensive smell. Sulphur is well indicated if, in the progress of the disease, the cartilages become inflamed and swollen. The patient has frequent weak, faint spells, with coldness of the extremities and even general chilliness of the body, and it has long been noticed that such patients are very liable to take cold. The 30th is here the most effective preparation, but I have used Boericke and Tafel's 200th with fine results.

Calcarea carbonica is classed by Jahr—in company with Sulphur and Silicia—as being one of the most reliable remedies for chronic coryza. The fore part of the nose is red, inflamed and swollen; The nose is dry and of very offensive smell; the nostrils are sore and ulcerated; the discharge may be thick and pus-like, or thin and watery. The mucous membrane is frequently moist during the day and dry at night. The little patient has a tendency to enlargement of the glands, and profuse sweat is often present, especially about the head and feet. Patients for whom Calcarea carbonica is suitable are very susceptible to external influences, as currents of air, cold, heat, noise and excitement. It is an additional indication when the catarrhal irritation extends from the nostrils to the air passages; hoarseness is a leading indication. "No remedy will be more frequently needed in irritations and sub-acute inflammations of the mucous membranes. Even in catarrhs which run on into structural degradation, simulating phthisis, it has proved to be the curative remedy, and the question may be raised if it will not arrest phthisis. A good remedy in scrofulous ozæna." (Brigham.) I have almost invariably used the orthodox 30th dilution, but, as Hughes well remarks, " the 3d is undoubtedly efficacious."

Silicia is one of the invaluable remedies with which we combat the deep-seated, morbid processes which occasionally attack the bones of the nose, even extending to the cribriform plate of the ethmoid bone. The tip of the nose is sensitive to contact ; the mucous membrane is excoriated and covered with crusts ; ulcers are found high up in the nostrils. It has been found useful in inveterate, dry coryzas, also in chronic ulceration of the Schneiderian membrane, with discharge of acrid water which makes the inner nose sore and bleeding. In Silicia the coryza is dry oftener than fluent ; the contrary is the case with the coryzas of Aurum, Alumina, Arsenicum, Asafœtida, and Baryta carbonica. "The perspiration on the head is more in Silicia than Calcarea, and if covered lightly soon becomes warm ; sweats more often towards morning." (Brigham.) I have never used Silicia lower than the 12th and have had excellent results from the 20th and from Boericke and Tafel's 200th dilution.

Kali bichromicum is a principal remedy for catarrhal inflammations involving nearly the entire respiratory tract, as well as the nasal passages. It has a very wide range of action, and has probably been less administered, at least on this side of the Atlantic, than it deserves. "It is one of the few drugs beneficial in caries of the bones of the nose, and useful in combating the constitutional effects of syphilis, when complicated with catarrhal affections of the nose and throat." (Morse.) The Kali bichromicum catarrh usually begins with a profuse mucous discharge, which at first is clear as water, but as the disease progresses the discharge is thick, tough mucus, which finally, on drying, fills the nose with hard, elastic plugs. Great pain is caused by the removal of these hardened masses, and they leave the nose very sore. There is a great accumulation of tenacious, ropy mucus, which is so viscid that it may be drawn out like a long thread, and the pathological state appears to be chronic ulceration of the nasal mucous membrane extending to the frontal sinuses, causing violent headache. "It produces deep and extensive ulceration ; the process carried on mostly in the cartilages, hardly producing caries of the bones. It is almost a specific for perforating ulcers of the septum, and many cases of cure are on record." (T. F. Allen.) The nostrils and upper lips are excoriated, with sore and swollen alæ, and the smell from the nostrils and mouth is very fetid. Kali bichromicum acts best on fat, light-haired people, and an additional indication is a concurrent affection of the digestive mucous membrane, indicated by foul tongue, eructations, nausea, and so forth. As to the dose, my experience exactly agrees with that of Dr. Hughes: " I recommend by way of dose the first six dilutions. The 3d is most commonly used, except in syphilis, where the lowest potencies of this salt and of the neutral chromate have been employed with most benefit. In acute affections, however, I nearly always prefer the 6th, unless I give the 12th."

Aurum metallicum is one of the chief remedies in chronic coryza, especially when the nasal bones are carious, as is often the case after abuse of Mercury and in syphilitic coryza. Still, as Dr. T. F. Allen remarks, it may also be called for in catarrh not yet involving the bones. The nose is swollen, red, inflamed and sore to the touch, especially the right nasal bone and adjoining parts of the upper jaw; there is a discharge of greenish yellow, offensive matter. The pains in the bones are aggravated at night, and they are accompanied by flow of tears. The nostrils are ulcerated, crusty, agglutenated, so as to impede respiration; ulcers in the nostrils covered with dry yellow crusts. The character-istic nasal secretion of Aurum is thick; in the sepia coryza the characteristic secretion is water. Dr. Morse remarks that when the scrofulous diathesis is marked Aurum muriaticum is preferable to Aurum metallicum. One would hardly look for the marked mental symptoms of Aurum in infants, though I have observed them in young children associated with chronic coryza. I have always used the triturations from the 6th decimal to the 12th centesimal and with excellent results.

Argentum nitricum is one of the leading remedies in both acute and chronic coryza, though Dr. T. F. Allen, an excel-lent authority, says that "the number of Nitrate of Silver catarrhs is not large." The coryza is at first dry; soon the mucous membrane becomes moist, and later a thick, yellow, purulent mucus issues from the nostrils. The alæ are painful and swollen, the nasal bones are painful, the septum is studded with bleeding pimples and the nose itches violently.

The scurfs in the nose become exceedingly painful ; if detached they bleed : bloody and purulent discharge in the open air, which stops in the house. The characteristic discharge is white and pus-like, mingled with clots of blood. The coryza is accompanied with constant chilliness, sickly look, lachrymation, sneezing, and violent, stupefying headache. The eyes and air passages are so frequently involved that Argentum nitricum does little good if the nasal passages alone are affected. Argentum nitricum acts best in the dilutions from the 12th to the 30th centessimal.

Sepia is a useful remedy in catarrhs arising from the retrocession of an eruption. The nose is inflamed and swollen, and the nostrils are angry and ulcerated with a painful eruption on the tip of the nose. There is obstruction of the nose with dry coryza and loss of smell. Dryness in the choanæ, though there is much. mucus in the mouth. Discharge of yellowish water from the nose, with cutting pains in the forehead. "This remedy is permanently indicated in cases where there is a discharge of green, bloody mucus from the nose, especially when accompanied by external inflammation of the nose. It is curative, too, in cases where there is ulceration high up in the nasal fossæ, accompanied by loss of smell" (Morse). Dr. Hermann Gross remarks that in the Sepia coryza putrid, subjective odor predominates, while in the Sulphur coryza objective stench from the nose predominates. I have used Sepia as low as the 12th centesimal, but have had the finest results from the 30th.

Alumina is one of the first remedies to be considered in chronic dry coryza when the mucous membranes, both nasal and aural, are broken down by ulceration, especially in scrofulous children. Such children are often chlorotic and prone to obstinate constipation. The nose is red, swollen and painful to the touch ; the nostrils are sore and scurfy; and the nasal mucous membrane is ulcerated, with discharge of a thick, yellowish mucus, or expulsion of yellowish-green scabs. The nose is stopped at night with dryness of the mouth, and the septum is swollen and painful to the touch. The throat is dry, especially on waking from sleep; the voice is thick and husky, and mucus accumulates in the posterior nares. Itching of the dorsum and of the alæ is an additional indication, and patients for whom Alumina is suitable take cold on the slightest exposure, yet feel better in the open air. I have never given Alumina lower than the 30th in this or in any other disease.

Baryta carbonica is useful for the chronic coryza of children with enlarged glands and large abdomens, weak both in body and mind. The nose and upper lip are swollen, and the nostrils are very dry with frequent sneezing. The Baryta carbonica coryza, however, is predominantly fluent—the direct opposite of Silicia—and the discharge is thick, yellowish and profuse. I have noted that Baryta carbonica is of little use unless the external nose is involved in the malady. This remedy has always been given by me in the 6th decimal trituration, and I have seen excellent results from it.

Lycopodium is one of the most reliable of our remedies for the dry form of chronic coryza, with much sneezing during the day; at night the nose is completely stopped, with dryness of the nose and burning headache. The nose is obstructed high up, with almost complete closure of the nostrils at night, so much so that the patient breathes with open mouth and protruding tongue. The morbid action frequently extends to the frontal sinuses, with frontal headache and thick, yellow discharge, which is at the same time, acrid and excoriating. The irritation is prone to extend to the air-passages, causing cough with loose expectoration, and the coryza then becomes somewhat fluent. This remedy is suitable for children who take cold easily, and who are troubled with derangement of the alimentary tract, and of this derangement the production of flatus is the most prominent symptom. Lycopodium is a most important agent when chronic coryza has extended to the air-passages, taking an ulcerative action and simulating pulmonary consumption. The remedy acts best in the 30th dilution, and seems to be of but little value below the 12th centesimal.

Lachesis is used in chronic coryza of the severest kind, syphilitic and mercurial as well as the still worse mercurio-syphilitic. The leading indication for Lachesis is a profuse watery running from the nose, accompanied by great soreness and swelling. The mucous membrane of the nose is swollen and bluish, and the nostrils are raw and bleed easily. The nose is full of scabs, and the discharge is pus mingled with blood, or there may be an extremely copious discharge of watery mucus. At the same time, the throat inclines to a low grade of inflammation resulting in plastic exudation, and the glands of the neck are swollen and tender. Epistaxis occasionally appears, and all the symptoms are worse in the afternoon and after sleeping. Lachesis acts well in all the dilutions from the 12th to the 200th; I have mostly used the 30th.

Graphites is suitable for children of lymphatic temperament who are subject to herpetic eruptions of the skin. Constipation is frequently present, and the patients easily take cold if exposed to a draught of air. Catarrh with obstruction of the nose; severe stuffed catarrh, with much nausea and headache, without vomiting; fluent coryza, with frequent sneezing, with pains in the sub-maxillary glands; heat in the forehead and face. Dryness of the nostrils, or alternate flowing and dryness; dry scabs with sore or cracked and ulcerated nostrils; bloody mucus from the nose, alternating with expulsion of dry scurfs; discharge of thick, fetid mucus. This remedy has usually been given in the 12th dilution, but I have had the best results from the 30th.

Kali carbonica is suitable for anæmic children of cachetic appearance, with puffy swelling over the upper eyelids, especially in the morning. Obstruction in the nasal passages, making it impossible to breathe through the nostrils when in a warm room; the patient, however, can breathe through the nostrils in the open air. The external nose is red and swollen with sore, crusty nostrils, or the nostrils are raw and bleeding. Fetid, yellow-green discharge from the nostrils; according to Hughes the characteristic discharge, is profuse and thin. The 30th dilution is most used, though good cures have been effected with the 12th centesimal dilution.

Kali hydriodicum is a principal remedy in chronic coryza of the nasal passages and frontal sinuses when occurring in syphilitic children, or in those poisoned with mercury. Ulceration of the internal nose, involving the frontal sinuses and antrum highmore; the nose is red and swollen with constant discharge of acrid, watery, colorless liquid, with violent lachrymation; anxious expression and restlessness; discharge of burning, corroding matter from the nose; the inflammation extends into the eyes and there is much conjunctivitis; the characteristic discharge is copious and watery, but it does not excoriate. I have always given this remedy in material doses, never higher than the 3d decimal dilution.

Nitric Acid is, according to Dr. T. F. Allen, a very potent remedy in syphilitic catarrhs of the nose and throat, also when such cases are complicated with mercurial poisoning. "I have derived more real satisfaction in seeing the prompt and lasting effects of this drug, not only in syphilitic catarrhs but lichen, ulcers, glandular affections, falling of the hair, etc., etc., than from anyother remedy. I think it is oftener indicated than any other, especially before the bones become much affected. I have occasion to use it every day in dispensary practice, and invariably the report is great improvement. The malar bones become sore and painful; soreness and bleeding of the inner nose; the nostrils are ulcerated, blood and bloody matter are blown out of them, with unpleasant smell. Nasal mucus goes down into the throat; inflamed and swollen alæ nasi, acrid matter from the nose at night; discharge of thick nasal mucus, corroding the nostrils; severe catarrh, with swelling of the upper lips and especially, night cough; stuffed catarrh with dryness of the throat on empty swallowing. I have never used Nitric acid higher than the 12th centesimal, though it would certainly act well in much higher dilutions.

Cyclamen Europæum is highly recommended in chronic coryza when the patient sneezes a good deal with profuse discharge, and rheumatic pains in the head and ears. "I had a fine illustration of the curative powers of Cyclamen in such cases with my colleague, Malaise, in Liege; the patient was a lady of upwards of sixty years old, and had been suffering from catarrh for years; it disappeared in less than twenty-four hours, to the astonishment of everybody" (Jahr).

Other remedies are Hepar sulphuris in scrofulous cases where there is great sensitiveness and the patient is chilled by the slightest draft of air—also in cases in which Mercury has been abused; the nose is swollen and painful, like a boil, and the nasal bones are painful to the touch, the discharges are thick and pus-like, and sometimes tinged with blood. Iodium for chronic coryza in cachectic, emaciated children of scrofulous habit with enlarged and indurated glands; the nose is painful and swollen, with fetid secretions which at times become a clear and continuous stream. Mercurius iodatus for syphilitic and scrofulous children with induration and swelling of the glands; the nasal bones are inflamed and the nostrils are sore and crusty; the nasal discharge is a tough, white or yellowish mucus which forms mostly about the posterior nares and adjoining parts; profuse, acrid, long-lasting discharges which excoriate the nostrils and upper lip Arsenicum iodatum, when the little patient has the tubercular diathesis with alternate chills and heat of the body; discharge of very irritating, watery mucus, corrosive and copious; at times this discharge is scanty and thick, sometimes it is tenacious and frothy. Stannum metallicum for severe catarrh with copious expectoration of thick, gray-green mucus, mixed with blood. Antimonium crudum, when the external nose is sore and painful, the nostrils angry, puffy and crusty, with a discharge of thick yellow mucus. Hydrastis for ozæna with bloody, purulent discharge, or chronic coryza with thick, tenacious secretions, more from the posterior nares, dropping down into the throat. Ailanthus for coryza, with rawness inside the nostrils; chronic nasal catarrh, with difficult breathing through the nostrils; the whole nose and upper lip covered with very thick, grayish-brown scabs. Asafœtida for pains in the bones of the nose, with a greenish offensive discharge, worse at night. Berberis, for chronic coryza of the left side, extending into the antrum of Highmore, with purulent yellow or greenish discharges.

Additional remedies are Ammonium carbonium, Ammonium muriaticum, Natrum carbonium, Natrum muriaticum and Magnesia muriatica.

APHORISMS.

1. Chronic coryza of infants is common, but it is exceedingly difficult of cure.

2. The best means of preventing chronic coryza is to attend to all cases of acute coryza, even those which seem to be insignificant.

3. Chronic coryza has little or no tendency to cure itself, the earlier the patient is attended to, the more rapid will be the cure.

CHAPTER IV.

Spasm of the Glottis.

Few diseases have received so many names as spasm of the glottis, and in this case, as in many others, "words without knowledge darken counsel," and, to use the forcible language of John Fletcher, "the alliance between nosology and nonsense is too palpable to escape the meanest capacity." It has been called "inward fits" by the vulgar, as if the word "inward" conveyed the slightest idea of the locality of the disease, though Trosseau and Pidoux, who style it "internal convulsions," almost sanction the name. With equal incorrectness it has been styled "goitre of infants," "suffocative catarrh" and "laryngeal asthma." Millar, who claims to have been its first observer, calls it after himself, and Kopp, who claims the same distinguished honor, does the same thing. Boerhaave styles it "asthma puerorum;" Hufeland calls it "catalepsis pulmonum," and Pagenstecher "asthma dentientium." It has been called "false croup," "cerebral croup" and "spasmodic croup," and Dr. Marshall Hall calls it "croup-like convulsions." Bouchut names it "phrenoglottism," Eberle makes one smile with "carpo-pedal spasms," and Mason Good, whose system of nomenclature is incomparably the most complicated we possess, calls it "laryngismus stridulus," which, however, seems likely to be the classic title of the disease. It has no affinity to asthma, croup or catarrh, and I shall use the familiar name of spasm of the glottis which conveys some definite idea to the mind, and which is sanctioned by some of the best of the French writers who speak of "spasme de la glotte."

Spasm of the glottis may be defined to be a spasmodic

contraction of the muscles which narrow the glottis—namely, the two thyro-arytenoid, two lateral crico-arytenoid, and the arytænoideus muscles—and this narrowing of the glottis is accompanied in very severe cases by spasmodic action of the diaphragm and intercostal muscles. As a result there is a succession of crowing, stridulous inspirations with a feeling of suffocation in the larynx, commencing suddenly, lasting at first for a brief period, and ceasing suddenly, usually with a fit of crying. The attack is unaccompanied by cough or any other evidence of laryngeal or thoracic disease, and as the disease advances other convulsive symptoms appear— strabismus distortion of the face and general convulsions, and peculiar convulsions of the hands and feet mark the more advanced stages. Should the two *posterior* crico-arytenoid muscles be affected, the very first attack would necessarily result in complete cessation of respiration and consequent death.

The most discordant views have prevailed as to the nature of this disease. Etmuller, who wrote in 1697, speaks of the "suffocative convulsions of infants" arising either from spasm of the muscles closing the glottis or paralysis of those opening it. Richa and Verdries, in the beginning of the eighteenth century, thought it was a laryngeal cramp caused by swelling of the thymus gland. In 1769, John Millar wrote on the disease, but as he confounded it with catarrhal laryngitis, and possibly with diphtheritic croup, we merely gather that he writes of laryngeal diseases running their course with attacks of suffocation and often ending in death. In 1795, Wichmann defined the disease to be a non-inflammatory form of croup—a purely nervous affection—without alteration of the mucous membrane. In 1830, Kopp published his well-known work in which for the first time he endeavored to give an anatomical basis to the etiology of this disease. The cause, according to this writer, is always hypertrophy of the thymus gland compressing the nerves supplied to the larynx, and this view was very generally held by medical writers for a number of years. In 1836, Ley announced that the disease arose from the pressure of enlarged glands on the pneumogastric or recurrent nerves causing paralysis of the abductors of the larynx. Rilliet and Barthez, and indeed most of the French writers, vaguely describe it as being a "neurosis," while Valleix doubts the propriety of classing it as a distinct disease. In 1841, Marshall Hall, in his famous work, "The Nervous System," referred this disease in all cases to reflex causes. "It is excitation of the true spinal or excito-motory system. It originated in—

   I.—1. The *trifacial* in teething.
      2. The *pneumogastric* in over- or improperly-fed infants.
      3. The *spinal nerves* in constipation, intestinal disorder, or catharsis. These act through the medium of
  II.— The *spinal marrow*, and
 III.—1. The *inferior* or *recurrent laryngeal*, the constrictor of the larynx.
      2. The *intercostals* and diaphragmatic, the motors of respiration."

Two years later, Elsässer announced his notable discovery of the connection between rachitis and spasm of the glottis, though he erred in attributing the latter always to pressure on the brain when the child lay on its back. In 1852, Bednar published the results of thirty-nine post-mortem examinations of children who had suffered from enlarged thymus glands, of which number but fifteen had suffered from spasm of the glottis during life, concluding that the disease did not depend upon thymic hypertrophy. In 1858, Bednar's observations were strongly confirmed by Friedleben, in spite of which Abelin, in 1868, maintained the old view, that spasm of the glottis often has its origin in swelling of the thymus gland, professing to ground the opinion on his post-mortem examinations. Professor George B. Wood attributes it to a "general morbid excitability of the nervous system, directed especially to the muscles of the glottis," and Dr. P. W. Bird considers that "it is not an independent disease, but merely a collection of symptoms consequent on disturbance of the nervous system in general, and of the respiratory nerves in particular." Later, Sir Dominic Corrigan stated the opinion that the disease was caused by a material change in the cervical position of the spinal cord, and Dr. Charles West maintains that it is often caused by the irritation of teething.

Several distinct influences are concerned in the production of this disease, and upon a proper appreciation of these influences successful treatment will depend. Many cases, as Dr. John Clarke long ago pointed out, depend upon an irritation of the brain, and this irritation is most likely the result of a local congestion near the origin of the pneumogastric nerve. In support of this view which is fast gaining ground, we have the undoubted fact that many cases of spasm of the glottis are preceded by well-marked symptoms of cerebral disease; and in cases of disease of the medulla oblongata external pressure has been known to cause the disease. It is well to remember, however, that the morbid appearances seen after death are frequently not the *cause* of the spasm of the glottis, but the *result* of the sudden apnœa. Many cases depend upon a rachitic condition of the bones of the skull—the "craniotabes" of Elsässer. Sir William Jenner noticed that rickets existed in every case of spasm of the glottis that he saw, save only two cases, and in ninety-six cases of spasm of the glottis examined by Lederer, rachitic softening of the cranial bones existed in ninety-two. In this class of cases there are probably changes in the nutrition of the brain as one result of the rachitic dyscrasia, and the spasm of the glottis is caused by the reflex influence of the morbid change in the cerebral mass. Strumous disease of the cervical and bronchial glands may cause spasm of the abductors of the larynx, which is the essence of the disease under consideration, by obstructing the venous circulation in the neck, and thus giving rise to irritation of the brain, which is again reflected upon the laryngeal muscles. Again, some medical writers distinguish an acute and a chronic form of spasm of the glottis; the acute form comprising the cases in which the spasms recur frequently, and in which death by suffocation often occurs after a few paroxysms; the chronic, those which have few paroxysms at comparatively long intervals.

Spasm of the glottis is a disease of northern climates and of the winter season; and the mild air of summer is a most powerful adjuvant in the cure. Out of forty-one cases Henoch noticed thirteen in the month of March, and Dr. Gee confirms these observations, for of 65 cases observed, 58 were in the first half of the year and only 5 in the second six months. The following figures show the number of these cases occurring in each month: January, 3, February, 11, March, 7, April, 13, May, 16, June, 8 (total 58); July, 0, August, 1, September, 0, October, 1, November 1, December, 2 (total, 5). Gee and Flesch simultaneously advanced the theory that the increased susceptibility to the disease is to be attributed to the exalted nervous condition of the children, resulting from the long confinement indoors. It seems to be rare in France, for when Rilhet and Barther published their first edition they had seen but one case, and they were acquainted with but one other, published by Constant in the *Bulletin de Thérapeutique*; when they issued the second edition of their work they had seen only nine cases in all. It is more common in Germany, and still more common in Great Britain, for, during the twenty years from 1857 to 1876 inclusive, the Registrar-General reports 7,318 deaths under ten years of age, and 37 deaths from ten to seventy-five years of age. Dr. Copland says that he has had as many as three cases under treatment at the same time; Ley reports having met with over twenty cases; and Dr. Charles West mentions thirty-seven of which he has preserved some record. Dr. Marshall Hall observes that "within the short space of one month I have seen five cases of croup-like convulsions." Dr. Condie speaks of it as being common in the United States, while Dr. J. F. Meigs remarks: "I do not think it is a common disease in Philadelphia, though it is certainly not extremely rare, since I have seen four cases myself and know of the occurrence of two other cases that proved fatal, and of two cases of recovery." The writer, whose practice is largely among children, has treated twenty-eight cases and has heard of many more in the practice of his medical friends.

Spasm of the glottis is, as a general rule, a disease of the first dentition, though the writer lately had a case in which the patient was five years old, and Meigs and Pepper remark that they had one very rare case in which the patient was seven years of age. The English Registrar-General, however, a few years ago reported 3 cases in which the patient was no less than seventy-five years old. Vogel remarks that the disease makes its appearance with the eruption of the first tooth and disappears with that of the last, adding that it occurs much oftener with the cutting of the incisor teeth, in the first half year of life, than with that of the canine and molar teeth. Gerhard assigns for this disease the period between the fifth and twenty-fourth month, and he says it is very rare after dentition terminates. Romberg relates that one of his own children was attacked with violent spasm of the glottis on the second day after birth, but it only occurred in a single paroxysm and did not return; and this distinguished writer thinks that the chief proclivity to this disease is manifested from the sixth to the fourteenth month, children of three or four years being exceptions. Heines says that of 226 cases which he attended, 174 were in the first year of life and the remaining 52 between the second and third years. Rilliet and Barthez, whose experience was but limited, observed this disease almost exclusively in infants of the age of three weeks to a year and a half, and Flesch states that it is rare after the twenty-first month. Heffen remarks that the disease is rare before the close of the fourth month, and he thinks that the majority of cases occur between the age of four months and the close of the second year; but if the disease is not developed till during the second year and is disposed to be tedious, it may last for a longer or shorter period beyond the limit named. He has further noted that if it occurs after the close of the third year it is less intense than during the first years of life, and he instances a mild case lately seen in a boy eight years of age. Of thirty cases taken indifferently from the practice of Drs. Meigs and Pepper and from various authors, 13 were six months or less of age, 11 between six months and a year, 4 between one and two years of age, 1 of two and 1 of 4 years of age ; so that of these thirty cases four-fifths were under one year. In Morell McKenzie's 31 cases the ages at which the attacks occurred were as follows: from birth 1 case, at 4 months 1 case, at 5 months 6 cases, at 6 months 5 cases, at 7 months 7 cases, at 9 months 3 cases, at 10 months 1 case, at 11 month 2 cases, at fifteen months 3 cases, at seventeen months 1 case, and at 23 months 1 case. In 31 out of 37 cases observed by Dr. Charles West the disease occurred between the age of six months and two years ; and in 48 cases Dr. Gee found 1 at six months, 19 from six to twelve months, 16 from twelve to eighteen months, and 12 from eighteen months to two years. Henoch saw sixty-nine children with spasm of the glottis, and 39 were between the ninth and thirtieth months, and 22 between the second and ninth months, and Salathè saw 24 cases, 4 in newly-born infants, 9 in those of from one to six months old, 6 in those from six to twelve months old, 4 from one to three years, one in a child twelve years old. Wunderlich thinks that the chronic form mostly occurs between the fourth and tenth months, and the acute form from the age of eighteen months to nine years ; and Hérard has an almost unique experience, for all his patients were over two years of age, and two of them were between three and four years old. Of the writer's 28 cases, eleven were less than six months, sixteen between six months and a year, and one was five years.

The following tables compiled from the English Registrar-General's Reports by Dr. Morell Mackenzie, showing the number of deaths, for the twenty years from 1857 to 1876 inclusive, from the disease occurring at different ages, gives the most conclusive evidence as to the importance of age as a predisposing cause.

Analysis of the English Registrar-General's Reports on the Mortality from Spasm of the Glottis:

CHILDREN UNDER 10 YEARS OF AGE.

	Totals	Under 1 Year.	1.	2.	3.	4.	From 5 to 10 Years.
FEMALES	2,547	1,487	691	152	94	60	63
MALES	4,771	2,915	1,365	213	97	63	88
GRAND TOTAL	7,318	4,402	2,086	365	191	123	151

ADULTS.

	Totals	10	15	20	25	35	45	55	65	75
FEMALES	13	5	—	1	2	2	1	1	1	—
MALES	24	7	1	2	2	2	3	—	4	3
GRAND TOTAL	37	12	1	3	4	4	4	1	5	3

After *age* comes *sex* as a most influential, predisposing cause in this disease. The Registrar-General's Report just quoted gives 2,547 females against 4,771 males in the first table, and in the second, which speaks of persons from 10 to 75 years of age, the numbers were 13 females to 24 males. Of the 16 cases seen by Hérard and Rilhet and Bartlez, 12 occurred in boys and 4 in girls, while of 183 collected by Larent, in which the sex was noted, 125 occurred in boys and 58 in girls. In Steiner's 226 cases the relative proportion of the sexes was 150 boys to 76 girls, while Vogel in his 15 cases had 11 boys against 4 girls. Of the 28 cases under my own care, 19 were boys and 9 were girls. Of Mackenzie's 37 patients, 21 were boys and 16 girls, while in Dr. Gee's 48 cases, 34 were boys and 14 girls. Almost the only statistics in contradiction to these are furnished by Salathé, who found only eleven boys in twenty-four cases, while of 297 cases seen at the hospital for sick children, London, 166 were males, and 131 females, making a total of 177 males, and 144 females; but these observations are too isolated and the numbers too few to invalidate the very conclusive figures already given. Steffen gives the following table, which "alone amounts to a demonstration," though the Registrar-General's figures being larger are still more conclusive:

Herard	saw	16	cases,	12	being boys and	4	girls.
Larent	"	183	"	125	" " "	58	"
Steiner	"	226	"	150	" " "	76	"
Henoch	"	61	"	49	" " "	12	"
Werner		26	"	15	" " "	11	"
Hachmann	"	14	"	12	" "	2	"
Pagensticher	"	18		14	" "	4	"
Kopp	"	10		9	" "	1	"
		554		386		168	

The precise cause of this greater predisposition of the male sex to spasm of the glottis is still an unsolved mystery.

Still another predisposing cause of spasm of the glottis is supposed to be *heredity*. Romberg says that in one family he attended two children who labored under this complaint (one of whom died), after three other children of the same family had fallen victims to it; and Gerhardt reports a family of nine, all of whom suffered from spasm of the glottis, and seven of these died. Dr. Ley quotes four instances from various authors in which three children in each family had the disease, and Powell saw one family of thirteen children, all of whom had had attacks of this malady. Werner saw two cases each in four families, and three children in another family seized one after another, and two of the writer's children have had severe attacks, but neither died. But the most striking illustration of this phase of the disease is that given by Reid, in which, out of a family of thirteen children, ten died of the disease and only one escaped an attack.

Do these cases prove an actual hereditary descent from parent to child? Romberg, a great authority on nervous diseases, is quite certain that it does, for he says: "There can be no doubt of the existence of an hereditary disposition; in many families several and even all the children, though they may have been differently brought up, both as to residence and food are attacked with a spasm of the glottis." Bouchert remarks that "it is sometimes observed amongst children born of a delicate, excitable, or nervous mother; and what is a strong proof of the original disposition of this disease is its successive appearance amongst all the children of the same family;" and Vogel, after observing that "the hereditary character of spasm of the glottis is interesting," goes on to say that "the mothers of the children whom I have treated for this disease were all of a tolerably excitable nature, and often complicated the child's disease by indulging in their habitual hysterical outbreaks." Morell Mackenzie thinks that the many cases in which the disease has attacked large families do not really prove its actual hereditary descent, but that they "strongly point to consanguineous influence; and he points out that the apparently strong proof afforded by the cases of Gerhardt and Reid may all be explained on the supposition that in each instance all the children were exposed to the same antihygienic influences. He illustrates this view of the question by the following case: "A gentleman of slightly strumous organization married a healthy woman, and had two boys and two girls. They none of them suffered from laryngismus, but the influence of the father's constitution was shown in the children by enlarged cervical glands, hypertrophied tonsils and early decay of the teeth. The family grew up, all married and all had children. In two of the families one child had laryngismus, and in one family two children suffered from the disease, and in one family three children were affected. In all four families the children were slightly rickety." Steffen correctly points out that all the cases which have been adduced to prove a real hereditary predisposition to this disease do not prove a true descent from parent to child, but only that several children of a family suffered from it—which is a very different thing. So that till descent from parent to child is clearly proved, we must conclude that the spasm of the glottis is not hereditary.

The prevailing opinion of those authors who have devoted most time to the investigation of this disease is that children subject to it are mostly delicate and feeble, and that it affects most violently those of scrofulous and rachitic constitution. On the other hand it has often been observed in children of the most robust and vigorous constitution, and some of the writer's patients were pictures of health. The proportion of children who suffer from both spasm of the glottis and rickets is undoubtedly very large, and Steffen, writing from European observations, is probably correct in asserting that the healthy constitution of the body which presents favorable soil for the commencement of the disease consists in, by far, the larger number of cases in a predisposition to rachitis. Dr. Gee reports rickets present in 48 out of his 50 cases, all of which, however, occurred among the poor, in whom all the causes of rickets would most likely be in full operation; Flesth says that three-fourths of his cases were rickety, and of Mackenzie's 31 cases, all of which occurred in private practice; 17 were slightly rachitic, while 2 were markedly rachitic. Steffen asserts that in by far the larger number of cases, say at least nine-tenths, rickets give rise to spasm of the glottis. But one of the writer's 28 cases suffered in any degree from rickets, and that one case only to a very small extent; and it is quite certain that on the North American continent the co-existence of rickets and spasm of the glottis is much rarer than in Europe, where, in a very large proportion of the population, the causes of rachitis are more actively at work than they are on this more favored continent. Elsässer looked upon craniotabes, which only appears in well developed rachitis, as being almost always the cause of spasm of the glottis; but, though Steffen says that "spasm of the glottis may be expected when it (craniotabes) is present," the writer has never noted craniotabes and spasm of the glottis occurring in the same patient, though he has seen 28 cases of the latter in his own practice and many more in the practice of others. Again, he has seen a large number of cases of craniotabes, none of which had ever suffered from spasm of the glottis, so that he looks upon the connection between the two morbid states as being at least problematical. Condie thinks that "it is very certain that, after the most careful analysis of the observations on record in reference to rachital softening of the cranium, that in the majority of instances, spasm of the glottis occurs in cases where not a trace of craniotabes exists." Curiously enough, Steffen himself admits that spasm of the glottis "in no way depends on it (craniotabes), and does not necessarily follow;" and Mackenzie, while admitting that the two morbid states often co-exist, says that "it does not follow that rachitis is to be regarded as the cause of laryngismus." On the other hand, Vogel, a great authority, considers that the connection between craniotabes and spasm of the glottis has been "satisfactorily demonstrated" by Elsässer, but he differs from Elsässer as to the precise *modus operandi*. Elsässer held that the pressure of the pillow on the soft occiput was competent to cause spasm of the glottis, while Vogel contends that "not the softness and depressibility of the occiput *per se*, but their *effects*, should be regarded as the exciting causes, as the meninges may thereby degenerate into an abnormally congested condition."

More influential than craniotabes in the causation of spasm of the glottis is *rachitis of the bones of the thorax*. Children who are born or brought up in a small, or damp, or cold house, who live in close or unwholesome air, who are badly or insufficiently nourished, and, above all, who are deprived of sunshine, are apt to suffer from disorder of the processes of digestion and assimilation, and in these unfortunate children rachitis is developed with melancholy facility —even in those whose parents had not suffered from rachitis in their childhood. Steffen, in his ingenious explanation, points out that an abnormal irritability of the nervous system is one of the most marked features of spasm of the glottis. This abnormal irritability is greatly increased, if not entirely caused, by the rachitic state in the following manner: the lateral flattening of the walls of the thorax leads to a marked diminution of the capacity of the chest, and that, in its turn, leads to a more superficial respiration, and thence to increased frequency of respiration; this, of course, at once necessitates an increased activity of the heart, greater wear and tear of the system, and consequent irritation of the brain and entire nervous system. Suddenly then, in such children, spasm of the glottis occurs, and in these cases it is not so much the bones that are at fault as the deep-seated disturbance of nutrition and, above all, the greatly increased irritability of the nervous system, without which spasm of the glottis is unlikely to take place.

Many excellent observers held to the purely nervous nature of spasm of the glottis. Mason Good asserts that it is "purely and idiopathically nervous," and Gregory says "it is caused by a high degree of irritability in the nervous system of the child." Felix von Niemeyer holds that "it depends upon a morbid excitement of the nerves by means of which contraction of the muscles of the glottis is effected," adding "that by uniform shortening of all the muscles at once, the vocal chords become tightly stretched, and the glottis is closed." But he considers that this irritation may be due to pressure along some part of the course of one of these nerves, or to centric irritation of the root of the vagus. Scrofula of the tracheal and bronchial glands is present in a considerable proportion of cases of spasm of the glottis, and these swollen glands probably act by pressing on the recurrent nerves. If this pressure is continuous, a constant wheezing is present, but usually the pressure, depending on the amount of blood in the glands, is moderate and variable, so that respiration is sufficiently easy. In most of these glandular cases the bones of the thorax are affected at the same time, and here Steffen's ingenious explanation would hold good. Finally, all experienced observers will concur in the opinion of the lamented Felix von Niemeyer, "*In most cases the pathogeny of this disease is obscure.*"

The *exciting causes* are very various. In general terms it may be said that any force capable of acting upon the general morbid excitability of the nervous system will produce the disease in those predisposed to it. The irritation of teething stands in the front rank, though Morell Mackenzie thinks that the influence of teething in the causation of spasm of the glottis is "enormously over-rated," an opinion in which the author cannot concur. Spasm of the glottis is rarely the first manifestation of morbid dentition, it is usually preceded by irritation of the brain, or disorder of the alimentary canal. Sometimes a child has an attack of spasm of the glottis whenever it cuts a tooth, but the first of these attacks is usually the most severe. Disorders of digestion are also a frequent cause, and hence it often occurs in children fed by hand. Weaning, according to Romberg, appears to favor the development and continuation of the disease. Sometimes it depends upon habitual constipation, and it may be caused by the sudden suppression of the diarrhœa of dentition. No single agency occupies a more prominent position in the popular pathology than "worms," and this omnipresent cause is capable of exciting an attack of spasm of the glottis in those predisposed to it. The disease, again, may depend upon some deep-seated cerebral affection, and two of the writer's most severe cases were caused by congestion of the base of the brain. It may precede hydrocephalus, and its occurrence in a child who is not teething and who is free from disorders of the digestive system, is always a suspicious circumstance. The mere act of swallowing occasionally excites an attack, and Prof. G. B. Wood says that infants are sometimes attacked with it when tossed playfully in the air. The writer has seen it follow infantile emotions as fretting and fright, and in such cases the disease is very liable to recur.

The briefest and, at the same time, the most graphic account of the disease is that given by Dr. John Clarke in his "Commentaries on the Diseases of Children." "The child is suddenly seized with a spasmodic inspiration, consisting of distinct attempts to fill the chest, between each of which a squeaking noise is often heard. The eyes stare, and the child is evidently in great distress; the face and the extremities, if the paroxysm continue long, become purple; the head is thrown backward, and the spine is often bent as in opisthotonos, at length a strong expiration takes place, a fit of crying generally succeeds, and the child, evidently much exhausted, generally falls asleep."

Spasm of the glottis often appears suddenly and wholly without warning, though the little one has sometimes been drooping for a few days, has lost appetite, and has been fretful and peevish. The lighter attacks merely consist in crowing inspiration, and this excites little or no alarm. If the attack takes place during the day the little one becomes pale, throws itself backward, and moves the hands and feet uneasily. Suddenly the crowing inspiration appears, the eyes roll up in the head, and the thumbs are turned in but not clenched tightly. At once the child cries out, and the attack, which had lasted but a few moments, is over. The little one is cross for a while, but soon regains it equanimity. If these light attacks occur in the night time the child wakes up, has the attack and then falls asleep, and unless the mother chances to be awake the disease may go on unnoticed for quite a time.

In severer cases the first attack is apt to take place at night—though Steffen says this is an error—very often towards midnight when, after the first deep sleep has passed away, the child suddenly starts up with great difficulty of breathing, inspiration being accompanied by a shrill crowing noise, which some observers compare to that of croup, but which really differs very much from it. The patient becomes much alarmed, and indeed the paroxysm is of the most urgent nature, and real danger is present. A few crowing inspirations take place, or, more rarely, some very laborious and audible expirations resembling a paroxysm of emphysematous breathing. Suddenly a more or less complete closure of the glottis takes place, the crowing sound ceases, the respiratory movements of the chest are arrested, and the thorax, the diaphragm, and even the anterior abdominal muscles become fixed and immovable. The crowing inspirations which precede the glottis seizure are usually accompanied by a flushed countenance, but the face now becomes pale and livid, and, if the paroxysm lasts long, this deepens into a cyanotic hue. The child throws its head back, the eyes roll in the head or stare straight forward, the great vessels of the neck become turgid, the mouth opens and the nostrils dilate, and a cold sweat suffuses the forehead and even the entire head.

In many instances general convulsions appear, especially if the paroxysm is severe and of long duration. All the muscles of the arms and legs are affected, the hands are tightly closed and the thumbs pressed into the palm, and at times even the wrists are bent inwards. Sometimes the hands are tumefied and reddened, and in almost all cases pain is caused by an attempt at extension. The spasm also affects the feet, the great toe is drawn apart from the other toes which are bent inwards, and the foot is rigidly extended, or, as in a recent case of the writer's, fixed in the form of talipes varus. These so-called "carpo-pedal" contractions are most likely accompanied by great pain. The general convulsions, even the episthotonos, evidently depend upon the convulsions of the glottis, for they appear and disappear with them, and the more exquisite forms partake of the character of epilepsy. Frequently the fœces, less frequently the urine, are passed involuntarily during the attack. The paroxysm terminates with one or more whistling inspirations, and the respiration, at first very irregular, assumes its accustomed rhythm, consciousness returns, the action of the heart becomes stronger and more regular, the cyanotic hue of the face gives place to pallor which, in its turn, gives way to the normal color, and the child is itself again.

In some cases, happily rare, the paroxysm assumes the form of a sudden spasm, almost without sound, which does not relax till the child is dead. These are the cases in which the posterior crico-arytenoid muscles are most likely affected.

Morell Mackenzie says that "the first attack of laryngismus often comes on at night—frequently towards eleven or twelve o'clock;" but Steffen asserts that "the supposition that spasm of the glottis has a special predeliction for the night season is an error held by very many." Amongst that many must be included the present writer, for a very large majority of his cases had the first attack in the night season, and most of the subsequent paroxysms occurred during the night. An attack is very short, say from five seconds to two minutes, though the extreme danger makes the time seem much longer, and attacks said to last half an hour will be found to be composed of a succession of paroxysms, with very brief intermissions, just sufficient to throw a re-inforcement of oxygen into the blood. The light attacks are usually short, the severe attacks are usually long, and the paroxysms show a strong inclination towards progressive severity in regard to intensity, duration and recurrence, and, consequently, danger. When the symptoms are of the character described as belonging to the severer variety of the disease, the paroxysm is almost certain to be followed by others in increasingly rapid succession, and the child may die almost at once. I had one case in which a fine, healthy boy of twenty months had the first and only paroxysm in the morning after his bath; the glottis closed almost without sound, and the little one died in less than a minute. On the other hand, I had two cases in which the little ones had but one single paroxysm of great severity, followed by perfect recovery, without any return of the disease. Still, as Romberg long ago noticed, it is only in very rare cases that recovery or death takes place during the first days of the illness. The duration of the affection depends very much on the exciting cause. It is rare that a child has one single attack; generally several paroxysms follow each other in rapid succession, after which the disease may disappear in consequence of the cutting of some teeth, or as the result of treatment. The first attack is usually the most severe, though when a second paroxysm rapidly follows the first one, almost before the child has recovered from it, the likelihood is that the second will be both long and severe. A strong child, previously in good health, may withstand several scores of paroxysms, provided they are not all of the more severe type and do not come in too rapid succession. Thus Dr. Benedict, of Philadelphia, reports a case, with the characteristic spasm of the hands and feet, which lasted for *four months and a half*, and was followed by perfect recovery.

As the child grows older the predisposition to the disease declines, and though the paroxysms may still recur, they are not nearly so severe, and the danger to life is evidently diminished. The explanation is that the laryngeal cartilages are firmer, the larynx is larger and especially wider, and the entire nervous system is less irritable and impressible. When the paroxysms do appear they merely consist in a feeling of suffocation and slight difficulty in swallowing, with slightly irregular respiration, but no crowing or whistling inspiration. Carpo-pedal convulsions have never been observed in children over five years of age, and they are somewhat rare from the end of the third to the end of the fifth year.

During the paroxysm it is a matter of some difficulty even to feel the pulse or ausculate the heart, and it is still more difficult to make thermometric observations. I have been unable to find any such observations in the libraries to which I have access, and can give but a few made by myself under exceptional circumstances. Alluding to this point, Vogel remarks " temperature of the extremities is much more likely to be diminished than increased," and Steffen says that " in view of the overloading of the venous system the body, and especially the extremities are, as a rule, cool and livid," but neither of these skilled observers appears to have used the thermometer. In mild cases, then, the temperature varies from 98° F. to 98.4 F., that is, a trifle below the temperature of the same child when in health, just what one would expect in a disease which is not only non-febrile, but, from the loading of the venous system, positively shows a temperature below the healthy standard. In severe cases of spasm of the glottis in children not suffering from any febrile disease, the thermometer placed in the axilla showed a temperature of 97.5° F. shortly after the commencement of the spasm, and as it advanced the temperature was lowered till it fell to 96.5° F. In one instance I succeeded in placing a very strong New York thermometer in the hand of a child just before the carpo-pedal spasm clenched the fingers, and it showed a temperature of 96.5° F. at a time when the axillary temperature was 97.5° F. nearly. I was unable to make observations on the feet in any of these cases, but I have no doubt but that they were cooler than the hands. I made careful observations in two children who were suffering from the fever of dentition at the time that they were attacked with spasm of the glottis, with the following results: The thermometer, which had shown a temperature of 100° F. on the previous day and 100.2° F. shortly before the attack, showed a temperature of 98.5° F. soon after the commencement of the attack of spasm of the glottis, and at its height the temperature was 97.5° F., showing in both cases an average of 1° F. above those children who were not suffering from teething fever at the time of the paroxysm of spasm of the glottis. Observations made by auscultating the larynx are still a desideratum, but I have been so intent on making thermometric observations that I have only made a very few in auscultation, and these are too few and too imperfect for publication.

Notwithstanding all that has been written as to the thymic origin of this disease, no characteristic lesion can be discovered after death, and the gland is sometimes increased in size, at other times it is smaller than usual, or it may be almost entirely absorbed. But hypertrophy of the thymus is by no means common, and chronic inflammation is quite rare. No alteration whatever can be discovered in the laryngeal nerves, nor in the laryngeal structure of the muscles; while the mucous membrane is slightly reddened only in the rare cases in which the child suffered from laryngeal catarrh as well as spasm of the glottis, so that, as far as the larynx is concerned, the purely neurotic nature of the malady is amply confirmed by *post-mortem* examinations. Craniotabes is present in many cases, though Steffen says this is by no means the rule, and next in frequency you meet with rachitis of the ribs. The tracheal and bronchial glands are frequently affected, but the lesions are not characteristic. Frequently they are more or less swollen and caseated, and this is particularly noticeable in the bronchial glands which are sometimes found to be mere collections of cheesy tubercles. At times, solitary glandular indurations may be found in the intestines, and in such cases tuberculosis is apt to be present in the lungs. Congestion of the brain and of its membranes have often been noticed, but these, though at times predisposing causes of the spasm of the glottis, are quite as often effects of it, and this is especially the case with the frequently-observed œdema of the brain. Steffen remarks that the softening of the medulla oblongata has, very occasionally, been demonstrated, and at times the brain has been found inflamed or even softened. Some observers have noted that the pneumo-gastric nerve was hardened, others have seen it softened.

Congestion of the lungs with engorgement of the right side of the heart is common, doubtless due to the asphyxia, and œdema of the lungs is very often present. Emphysema of the lungs is also found as a result of the irregular and spasmodic respiration, but, as Dr. J. Lewis Smith points out, "slight emphysema occurring in the upper lobes is common in infants, even those who have had no serious disease of the respiratory organs." The blood is darker than usual, and Bednar, and also Rilliet and Barthez found the heart and great thoracic vessels filled with black fluid blood, though Loeschner asserts that he has always found the thoracic organs somewhat anæmic. Finally, more or less congestion of the intestinal mucous membrane is usually present, and indeed any organ of the body may be congested as a *result* of spasm of the glottis.

Attention to Dr. Cheyne's pathognomonic sign will tend to prevent errors in diagnosis, "a crowing inspiration, with purple complexion, *not followed by cough.*" It has been confounded with croup, but in croup the difficulty of breathing is permanent or nearly so, and it affects expiration as well as inspiration; but in spasm of the glottis it is inspiration which is affected. In croup respiration is affected, though with difficulty; but in spasm of the glottis respiration is, for a brief time, positively stopped. Croup is accompanied by severe cough, but this symptom is wholly absent in spasm of the glottis. In croup the child is hoarse, but hoarseness is not an integral part of spasm of the glottis. Croup usually follows exposure to the cold; spasm of the glottis has little dependence on that cause of disease. Lastly, croup is usually accompanied by fever, and it has convulsions only when about to terminate fatally, while spasm of the glottis has no fever and has a most characteristic species of convulsion.

Spasm of the glottis may be confounded with œdema of the glottis, but the last mentioned developes gradually, whereas spasm of the glottis arises suddenly. The experienced finger can easily detect the hard and swollen cherry-like mucous membrane of the epiglottis, so characteristic of œdema, while nothing of the sort is found in spasm of the glottis.

There is a close resemblance between the sounds of whooping cough and spasm of the glottis, but in the latter disease there is no cough, no expectorations, no vomiting and no rattling of mucus in the lungs. Whooping cough almost always has a catarrhal stage lasting a week or ten days, while spasm of the glottis rarely has any prodromal stage whatever.

"Paralysis of the abductors—a very rare affection in childhood—might be mistaken for spasm of the adductors, and it is thus important to carefully distinguish between these two conditions. In the paralytic cases there is, as Dr. Marshall Hall has pointed out, "*a constant but partial closure*" of the glottis, the vocal cords never being abducted from their paralyzed position, but always leaving a small opening through which the air can pass. In spasm of the adductors, on the other hand, there is *inconstant but complete closure of the glottis;* in other words, there is considerable movement of the cords, which are at one moment widely separated and at another so closely approximated that air cannot pass through the glottis. The symptom in the one case is constant dyspnœa, increased on the slightest exertion, whilst in the other it is constant dyspnœa, with complete intermission between the attacks. This, however, is not an absolute law, for on three occasions I have seen slight constant stridor in the case of children in whom the other symptoms were of spasmodic character, carpo-pedal contractions and convulsions (Morell Mackenzie).

This is always a grave disease, for even in mild cases, serious symptoms may arise and the prognosis changes at once, and Condie remarks that ' a sudden and very severe paroxysm may unexpectedly occur at any moment, particularly during the period of dentition." Dr. John Clarke says that the patient rarely recovers, and Dr. Reid collated 289 cases showing a mortality of 115. Dr. Cheyne and Dr. Gooch both state that it proved fatal in one-third of those attacked, and only one of the 9 cases seen by Rilliet and Barthez recovered, and Hérard saved but one out of seven. Of Sir Henry Marsh's cases, 5 recovered and 2 died, and of Dr. Hersch's cases, 3 died out of 5—and one of the fatal cases was complicated with whooping cough. Bouchut thinks that rather more then than one-half die, while Lorent notes 45 deaths of 100 cases occurring in boys and 32 deaths in 100 cases occurring in girls. Wunderlich says that one-third of all those attacked, and the majority of those visited with severe attacks die, and Steiner, a physician of vast experience, is sure that the great majority die.

On the other hand Steffen says that, if we include the lightest cases, which we certainly should, the prognosis is, on the whole, favorable, and Salathé lost but 2 cases out of 24.

Morell Mackenzie says that spasm of the glottis "rarely proves fatal," which is not by any means the general experience. Notwithstanding all these opinions, almost all of them unfavorable, the writer's 28 cases all recovered save two, one of which was the *foudroyante* case referred to, and the same result may almost always be expected from an enlightened homœopathic treatment.

Spasm of the glottis is more dangerous to young children than to older ones, for in the latter the laryngeal cartilages are harder, the larynx is wider, and the nervous system is less impressible. Children at the breast do better than those who have been weaned. The prognosis is more favorable in girls than in boys. Emaciated children have less chance of doing well than well-nourished children. The lower the temperature during the attack the greater the danger. The longer the interval between the spasms the better the chance of recovery. General convulsions or carpo-pedal contractions add greatly to the gloom of the prognosis. The danger is greater when the malady results from intracranial disease than when it depends on dentition or on some stomach attack. Diarrhœa, continued vomiting, or anything which lowers the powers of life, vitiate the prognosis. Ryland and Ley both refer to bronchitis as an occasional exciting cause of the disease, while Steffen asserts that "acute pneumonia usually effects a material abatement and even complete disappearance of the laryngeal cramp." Children far gone in rachitis, especially if craniotabes is present, are usually considered less likely to recover than those of healthy constitution, yet Steffen maintains that "the most favorable cases are those in which it is developed as the result of rachitis." Scrofula, which finds its expression in swelling and caseation of the tracheal and bronchial glands, renders the prognosis more doubtful. Children who have had one attack should never be considered safe till the completion of the first dentition. Long-continued and severe spasms, with cyanotic face and symptoms of suffocation, often presage an unfavorable issue. If remedies are promptly and faithfully used by a physician who has carefully studied the disease the chances of recovery are much better than if remedies are carelessly used by a practitioner who knows little or nothing of the disease.

Death takes place in three distinct modes. The first is apnœa, when the child is choked during the paroxysm; first respiration is suspended, then after a few hurried beats the pulse ceases, and the lungs and heart are found to be flooded with dark blood. Or death may take place when the current of blood is prevented from passing from the brain to the heart and lungs, and if this state—really congestion—is not promptly relieved, effusion takes place and the child dies comatose. Lastly, death may take place from exhaustion; the strength is reduced by a constant succession of severe paroxysms, diarrhœa sets in, and death closes the scene.

Sambucus is the classic remedy of our school for spasm of the glottis, recommended before all others, though it cannot be looked upon as being the leading specific, and there is a growing inclination to question its efficacy. Dr. Drysdale has not seen any good effects from it, while Dr. W. S. Searle, of Brooklyn, after stating that Sambucus was the remedy selected by Hahnemann for the disease, goes on to say that "its signal failure to cure the large majority of cases has led some to question whether the usual sagacity of the master did not desert him on this occasion;" adding further, "the fact, however, does not prove that the remedy is not homœopathic to some modalities of the disease, and the trouble lies in our failure to discover these modalities; not in Hahnemann nor in the remedy." Bæhr thinks that the striking case reported by Hartmann does not represent a high degree of spasm of the glottis, while Farrington thinks that "its symptoms do not seem to point distinctively to a spasm of the glottis." Ruddock recommends during the attack a very prompt administration of Aconite, alternated with Sambucus, for "fear of suffocation and dry cough," but as the indications for Aconite are wholly distinct from those for Sambucus, it is decidedly best to give the remedy indicated, *singly and alone*. Hughes, one of our best authorities, says that "Sambucus is in high esteem."

The attack takes place suddenly; the patient awakes from a kind of lethargy with the eyes and the mouth open; he raises himself in bed with great anxiety and dyspnoea, the respiration is oppressed, with wheezing in the chest, the head and hands are puffed and bloated, with dry heat over the whole body, no thirst, small, irregular and intermittent pulse. The patient tosses about anxiously and is unable to sleep. There is no cough, and the attack principally occurs from midnight to 4 A. M. After the paroxysm, the child perspires profusely, and Sambucus is said to be the leading remedy when the disease originates in suppressed perspiration. Dunham points out that while Chlorine has difficulty in expiring, none in inspiration, Sambucus has the reverse. Searle gives the following determining indications: " Burning, red, hot face, hot body, with cold hands and feet *during sleep*. *On awaking*, the face breaks out into a profuse perspiration, which extends over the body, and continues, more or less, during the waking hours; then, on going to sleep again, the dry heat returns;" adding, "should you ever meet with a case presenting these peculiarites, you may be sure Sambucus will cure it, and nearly as sure that it will fail if these symptoms are absent." During the last few years I have only met with two cases in which Sambucus was indicated, but the results were striking and prompt, and I was greatly helped by Searle's laconic but invaluable indications. Sambucus has usually been given in drop doses of the 1st, 2d or 3d dilutions, though Hartmann thinks that a "higher attenuation may perhaps do better and prevent a recurrence."

Moschus is recommended by several of the standard writers, though Hempel says that it acts in accordance with the principle of Contraria, while Hartmann thinks that it is very far from being a specific remedy. " Moschus is variously recommended for this disease, but we cannot see its homœopathicity to it. We are not acquainted with any decided cures that Moschus has effected in this disease, Hartmann's statement to the contrary notwithstanding."

(Bæhr.) Drysdale brackets Moschus with Sambucus, but he has never seen any good effect from either. Searle states that it "is said to have cured laryngismus, probably hysteric in character," while Hughes affirms that his own experience has led him to believe smelling at Moschus to be the best means of relief during the paroxysm. "Moschus causes a spasm of the throat, larynx and lungs—sudden sensation, as if the larynx closed on the breath, as from inhaling sulphur-vapor. It is more applicable to hysterical cases, and possibly to spasm of the glottis during the course of diseases which exhibit impending paralysis of the pneumogastric nerve." (Farrington.) Dr. Pemerl, of Munich, writes: "If neither the chest nor the abdomen is affected, and we have only to battle against a reflex irritation of the nervous trigeminus, I prefer the exhibition of Moschus 1 or 2 in quickly succeeding doses, and usually I could be satisfied with the results; but Moschus is *never* of any use whatever when abdominal or thoracic affections underlie the spasmus glottidis." Kafka reports a case of laryngismus spasmodicus, in a girl five years of age, of light complexion, with sudden attacks of crowing and protracted inspiration. The spasm of the glottis was improved by Veratrum, but cured very quickly with Moschus 12, one drop every two hours. Laurie recommends one globule of the 3d dilution in a teaspoonful of water, repeated at intervals of half an hour until three doses have been given; and subsequently at intervals of two hours, until decided amelioration or change. For olfaction, as advised by Hughes, the 1st decimal trituration would be preferable.

Aconite is not even mentioned by either Hughes or Bæhr, and Searle omits it from the list of remedies—Chlorine, Mephitis, Sambucus, Moschus and Lachesis—of which he says, "besides these five, I am unable to find any remedies that have cured, or are likely to cure, a case of true spasmus glottidis." On the other hand, according to Prof. Walter Williamson, Aconite is "specific," while Hempel says "we have cured more than one spasm of the glottis radically with nothing but the first attenuation of *Aconite root*," and again he says, "we affirm that we have effected cures of this disease with Aconite alone, without using any other medicine. We mix a drop or two of the first decimal attenuation of the root in two tablespoonfuls of water, and give ten or fifteen drops of this mixture every two or three minutes, until the patient is decidedly easier. If no positive relief is obtained after giving a few doses, we substitute a drop of the tincture in the same manner as before." Jahr, under the caption of "Spasm of the Glottis, Asthma Thymicum, Kopp," writes, "After I had cured my daughter, a child of five years, in 1849, of this disease, which set in one morning with all the frightful symptoms of true croup, and I expected every moment to see her perish of asphyxia, in less than ten minutes, by means of a single dose of Acon., 30, three globules; I have commenced the treatment of this spasm in every subsequent case with Acon., thereby unfortunately creating a belief that croup can be wiped out, as it were, by a stroke of magic. Not every case of spasm of the glottis can be cured so easily with Acon. alone, although this remedy never fails, if no complications are present, to afford speedy help." I side with my lamented friend, Hempel, for I have seen the happiest results from Aconite, given as he directs, in spasm of the glottis. Aconite is indicated when a suffocating cough comes on suddenly at night with hoarse voice and shrill outcry; the respiration is short and anxious; the skin is dry and hot; the pulse is full, hard and greatly accelerated. Good results have been seen from all doses, from the Hempelite mother-tincture to the Hahnemannian 30th dilution, but in acute cases I have never given it higher than the 1st decimal dilution, while in protracted cases I have used from the 6th decimal trituration to the 12th centesimal dilution.

In 1869, writing on this disease, I spoke as follows on the use of Sanguinaria: "My own experience leads me to look upon Sanguinaria as being the Imperial Guard of all the remedies for spasm of the glottis. After using this remedy successfully in the various forms of croup, I was induced to give it in two apparently desperate cases of spasm of the glottis after the unsuccessful use of Aconite, Sambucus and other apparently well-indicated remedies. I was gratified to find its administration followed by rapid and durable cures, and I now look upon it as being the first remedy. I give it in the form of an acetous syrup." Now, after an enlarged experience, I feel still more confidence in this little-used but invaluable remedy.

Brilliant cures have been effected with Arsenicum album, especially when the disease assumes the chronic form. Drysdale mentions it as one of the medicines which he has found, on the whole, most useful in this disease, and Bæhr thinks that it deserves our attention if, as is often the case, the disease attacks feeble children with marked symptoms of cerebral anæmia. The attack is preceded for several days by catarrhal symptoms; the little patient goes to sleep quietly, and the spasm comes on suddenly in the night, threatening suffocation; the respiration is short and hissing. There is great anguish with copious perspiration; prostration of strength with aggravation of all the symptoms between midnight and daylight. The child breathes freely between the attacks, but is languid and restless. "In my opinion Arsenicum is the true specific for this disease. Not to mention the peculiar sense of suffocation or constriction in the larynx, with stoppage of breath, what drug has more than Arsenic the peculiarity of producing such a paroxysm at night, waking the child suddenly? or after trifling causes such as crying, laughing, getting choked by a little food or drink, etc.? What drug has the typical recurrence of the first paroxysms at decreasing intervals? the apparently insignificant prodromi of such a dangerous disease? the sudden disappearance of the spasm by violently shaking the child?" (Hartmann.) I have had better results from the 30th dilution than from the lower preparations, and I never saw any good results from Arsenicum save in cases such as Bæhr describes.

Belladonna is the principal remedy when the brain is seriously involved, when the head is hot and the face alternately flushed and pale, and all the symptoms point to cerebral congestion. It is of great service when the child's head is large and when the carpo-pedal spasms are present. Hughes says it must be given where there is arterial excitement and cerebral congestion, and Jahr recommends it especially in the case of scrofulous children. Bæhr observes that "Belladonna is only adapted to spasm of the glottis when occurring as a secondary symptom in other diseases," but the force of the objection is broken by the fact that spasm of the glottis is very often indeed "a secondary symptom in other diseases," but not the less dangerous on that account. The little one is over-susceptible to impressions, and this is aggravated by bright light, noises, the slightest contradiction, by dentition, or by the presence of irritating or indigestible substances in the intestinal canal. The respiration during sleep is irregular or intermittent, and on falling asleep the child wakes and starts as if frightened. The sleep is restless, tossing about the bed, quarrelling in sleep, talking or crying out. The larynx, which had been sensitive to pressure, suddenly feels constricted with violent struggles for breath, and the smallest quantity of fluid drank excites a spasm. The brain is excited, the face red, the eyes injected with squinting or dilated pupils, the teeth are clenched, and fearful convulsions of the flexor muscles set in; frequently opisthotonic convulsions. "It seems to be indicated by the characteristic spasm in the larynx, and may be given to children of every variety of constitution, plethoric or lymphatic, scrofulous, rickety, etc., and more particularly when bad management has induced collateral complications or cerebral affections." (Hartmann.) Belladonna has been used in all potencies in this disease. Thus Dr. E. Clarke, of Portland, reports a case in the *North American Journal of Homœopathy*, vol. XX, in which Belladonna in the 200th and 400th dilutions rendered excellent service, and on the other hand, Dr. J. N. Tilden, of Peekskill, N. Y., reports the following cases, which are really model cures, in the *Homœo-pathic Times* for 1878:

"CASE 1.—A delicate child, æt. 8 months, artificially fed, digestion in perfect condition. His paroxysms were always precipitated by crying from anger. They were characterized by a sudden and complete cessation of respiration, as if the rima glottidis were completely closed to the entrance of air, and accompanied by alarming lividity of the face, lasting for from ten to twenty seconds, when the first inspiration would be accompanied by a shrill crowing sound almost identical with the characteristic inspiration of whooping cough. After this prolonged inspiration the breathing would be irregular and sighing, and the discolored features would be followed by pallor, accompanied with great prostration, and cold perspiration lasting for half an hour or more. These alarming attacks occurred at irregular intervals, sometimes daily, often at longer periods. Strict attention to regimen, abundant out-door recreation was directed, and Belladonna 1st dec. given internally every two hours while awake. A marked diminution in the severity of the symptoms was at once noted, and after a few days' treatment the attacks ceased entirely.

"CASE 2.—Child, æt. 9 months, suffering from teething and indigestion, had paroxysms every time he waked from sleep. In this instance they consisted of ineffectual spasmodic efforts at inspiration, attended with the same shrill, crowing sound mentioned as occurring in the other case. This patient did not have so much congestion, nor were the paroxysms followed with so great prostration as in the previous patient; but during the attacks, which lasted one or two minutes, it seemed as if the little fellow must surely suffocate.

"The difference in the symptoms noted in the two cases was probably owing to the fact that in the first case the rima glottidis was entirely closed, and in the second, though rigid and unyielding, it was open sufficiently to allow the entrance of a limited amount of air.

"The treatment was the same in this case as in the preceding one—Belladonna—and the result was equally prompt and satisfactory. The paroxysms were at once ameliorated, and after three or four days there were no more symptoms of them."

My own experience with Belladonna in this disease decidedly leads me to favor the measure dose, though I have never descended to the 1st decimal dilution, having seen excellent results from the 3d to the 5th decimal. In the intervals of the paroxysms I have given the 12th and 30th centesimal dilutions with fine results.

Ruoff and Jahr speak favorably of Ipec. Searle does not mention it, and Pemerl, of Munich, recommends it "when we find sabilant rouchet with dry, titillating, frequent and tormenting cough—a morbid state little likely to be found in pure spasm of the glottis—and these indications can only hold good when a laryngeal catarrh precedes or accompanies the disease. "Ipecacuanha has been mentioned as a remedy; indeed the symptoms justify this recommendation. We should not, however, overlook the fact that asthmatic difficulties do not really occur in this disease. Relief is easily afforded if a remedy is given at the outset, but it does not last, and we cannot recommend a remedy as a specific, unless it controls the whole disease. We admit, however, that Ipecacuanha may have an excellent effect, for the time at least, in a catarrh accompanying spasm of the glottis." (Bæhr.)

I have given Ipecacuanha with success when the spasm of the glottis was excited by the reflex influence of indigestible food, the attack being preceded by nausea and vomiting and even by purging; also when the disease was caused by taking cold, and thus was associated with a catarrh of the larynx. The symptoms are rattling in the trachea and lungs from accumulated mucus, with spasmodic contraction in the laryngeal region, with threatening suffocation, anxious and short or sighing respiration, together with purple color of the face and cramps and rigidity of the body.

"As regards laryngismus stridulus, having had several

cases of it, we can speak from experience of the efficacy of Sambucus and Ipecacuanha in curing it. We once had a case of laryngismus in which the whooping-cough supervened. The suffocative spasms were of the most painful description. When a fit of coughing came on, and the child had at the same time an outburst of passion (the usual exciting cause of the attacks of laryngismus in this case), the child would scarcely have begun to cough when he lost all power of respiring altogether. He grew literally black in the face; his head fell backwards, and there he lay for some seconds apparently beyond all recovery. He revived on putting his head out of the window, or dashing cold water in his face. Ipecacuanha effected a cure." (*British Journal of Homœopathy*, XII, p. 457.)

As to the dose, Laurie advises to dissolve 3 globules of the 3d dilution in six teaspoonfuls of water; give a teaspoonful every quarter of an hour, until three doses have been given, after which the intervals must be lengthened, or the medicine suspended, if decided improvement or a cessation of the symptoms of impending suffocation ensues. I have always used the 3d, 4th, or 5th decimal triturations in the manner recommended by Laurie.

Gelsemium sempervirens was introduced to the profession by Dr. E. M. Hale, of Chicago, who considers that "if not a curative remedy, properly, it will be a valuable palliative, used in drop doses of the first dilution, or mother tincture, frequently repeated; it must procure relief in a majority of cases, while during the intermediate time it should be alternated with Belladonna, Hyosciamus, Arsenicum or Moschus." In a recent edition of his invaluable work he writes as follows: "The primary effect of Gelsemium is to paralyze the muscles of the tongue, glottis and the whole apparatus concerned in deglutition and vocal efforts, but this paralysis is not attended by the numbness and tingling caused by Aconite; the secondary effects of the drug result in spasmodic and tetanic conditions of the same muscular structures.

You will be able to cure as I have cured, some sudden and

alarming *paralytic affections of the throat* with the high dilutions; while in *spasm of the glottis and spasmodic croup* you will be successful with more material doses." Ruddock remarks that it is an excellent reserve medicine for an occasional acute attack which does not yield promptly and fully to Aconite, and it is one of the very few remedies endorsed by Raue. The indications are long inspiration with croupy sound, expiration sudden and forcible. Sudden and severe attacks of dyspnœa, with crowing noise, profuse perspiration and darkly flushed face.

In the fifth volume of the *Homœopathic World*, Dr. Adrien Stokes of Southport, England, relates two cases of spasm of the glottis cured by Gelsemium, of the first of which the following is a summary: The patient, twenty-one years of age, had had diphtheria, and three months afterwards, when first seen by the doctor, appeared to suffer from laryngeal catarrh. There was some tenderness of the trachea on touch, and an increasing difficulty of inspiration towards evening. Mercurius was given, but at midnight the doctor was summoned, and found the patient in bed, propped up with pillows, her hands pressed upon the bed beside her, the face ashy, the lips livid, and the countenance wearing an anxious expression. Respiration was very difficult, and the pulse thready and feeble. The finger-nails were livid, and the hands stiff. The larynx was very tender, and became increasingly so. Deglutition was difficult, and became more so as time went on. The doctor saw that he had before him a very formidable case of trachetis, with intense spasm of the glottis; and as the patient had only just recovered from a serious illness, he was inclined to fear this one would turn out to be rather unmanageable. Aconite and Belladonna were given in succession, but equally without benefit; the distress increasing with an increasing dyspnœa, the lips were blue and the hands also, the arms being rigid up to the elbow. In this emergency Dr. Harvey was called in, when Bromine was given by inhalation and Aconite every fifteen minutes, but the patient got rapidly worse, and the medical attendants were in fear of seeing her perish of asphyxia before their eyes; the power of deglutition had almost ceased—hands and feet were cold. In this desperate strait, Stokes happily thought of Gelsemium, of which a portion of a drop of the mother-tincture was at once given. The effect was something akin to the marvellous. "Scarcely had the fluid passed over the tongue when we saw the inspirations lengthen, and felt the hands relax from their rigidity. The countenance began at once to brighten, the hands soon regained a more natural appearance, and the whole bearing of the patient was easier and happier. In a quarter of an hour we gave the remains of the dose, and so rapid was the improvement that in another quarter of an hour we were able to go home to bed. I had been with her four hours, but had I thought earlier of using Gelsemium I need not have had an anxious and broken night." A rapid cure followed, though the larynx continued intensely tender and deglutition difficult for a week.

The second case was a young lady thirteen years of age, prone to laryngeal disease. "I went at once and saw this child in a state of intense distress. The old nurse was trying to hold her on her lap, but she was dashing herself about in a frenzy of fright and agony. The face was purple, the eyes protruding, the larynx was spasmodically jerked up and down, and suffocation seemed imminent. I thought at once of Gelsemium and how it had served me before. So mixing two drops of mother tincture in four teaspoonfuls of water, I gave her one teaspoonful of the mixture, and bade the mother watch the effect. In five minutes there seemed to be a slight improvement, inasmuch as the movements were less frantic and violent. A second dose was given, and in five minutes more a visible change had come on. The patient could now take breath more easily, sat still on the nurse's knee, and the acute and strident sound of the inspiration had given way. The mother watched and wondered; but we gave another dose at the expiry of five minutes after the last, and by that time all distress had passed away. I remained while the patient was being put to bed, and was in the house half an hour in all. The larynx remained very tender for a week, but I continued to treat the case with Gelseminum and Mercurius, and kept her in one room all the time."

Iodium is not merely homœopathic to the acute attack, but it is also one of the deeper-acting remedies which *must* be given if the cause is to be reached. We owe this remedy dy to Bæhr, who thus introduces it to the notice of our school: " Iodium is doubtless a very excellent simile, and is likewise adapted to all three above-named causal morbid conditions. (These conditions are rachitis, a deficient involution or hypertrophy of the thymus gland, and swelling of the bronchial glands.) With this remedy alone, given every other day at the fourth or sixth attenuation for four to eight weeks; we have cured five undoubted cases of spasm of the glottis, which evidenced their malignant nature by the fact that every subsequent attack was more violent than the preceding one. The patients were children not yet a year old, but only one of them showed an enlargement of the thymus gland. Supported by such striking curative results, we cannot be accused of hazardous speculation if we prefer this remedy to all others, as long as the general organism has not become too much reduced." The following are the indications usually given for this remedy, though it must be confessed that they are somewhat vague: Rachitic children; swelling of the bronchial glands; tightness and constriction about the larynx; soreness, hoarse voice, etc.; enlarged glands may even cause paralysis of the laryngeal, tracheal and bronchial nerves. Dunham gives the following indications: enlargement and induration of the glands, cervical and mesenteric; absence of appetite; utter indifference to food; scanty, high-colored urine; clayey evacuations; emaciation; yellow skin: action of the heart feeble, and much increased by motion. Guided by these indications, Dunham reports the following striking case:

" The case for which I recently prescribed Iodine was that

of an infant of ten months, whose mother states that, early in life, the child had marasmus, and was very low. Recovering from this, under homœopathic treatment, she had a wheezing or rattling in the chest, which gradually increased for two months, until she could be heard breathing at a great distance. She coughed for a week or two, then the cough ceased. About August 1st she began to have spasms of breathlessness, occurring usually at night, when asleep, and during the day while asleep, and seeming as if they would take her life. I could not distinguish a special indication for any remedy in any peculiarity of these spasms, and otherwise the child seemed perfectly well. I therefore adopted a plan which has often helped me in blind cases. I went back on the line of development of the child's symtoms, until I found symptoms which furnished an indication and then prescribed as though these symptoms were now present. Adopting this plan, it may be remembered I, years ago, prescribed for a deaf young man of seventeen years, Mezereum, which corresponded to the milk crust, the suppression of which, twelve years before, had been immediately followed by the deafness. I prescribed just as though the milk crust was actually present, and the deafness was speedily and permanently cured. Acting on this plan, I recalled the marasmus which the child had had, and the symptoms of which, as described to me, indicated Iodium. This remedy was certainly not contra-indicated by the affection of the glottis, which was, I think, a partial paralysis and not spasm. The attacks of dyspnœa gradually ceased, and within ten days had disappeared. The potency used was the two hundredth.

Dr. Dunham says that the breathing furnished no particular indication for Iodine more than for Spongia or Sambucus, so that we must fall back on the general indications already given. Personally, I have had no experience with this remedy.

Chlorine is, according to Searle, the most prominent of the remedies for spasm of the glottis, and he thinks that a large majority of instances of simple idiopathic spasmus glottidis may be expected to yield to it. Still, he candidly admits that the symptoms produced by it are those which occur in every case of spasm of the glottis, to a greater or less extent, adding that "the characteristic and distinctive symptoms of the drug have never been evolved." The attacks come on suddenly and without warning, the child takes a long inspiration with a slight crowing noise, but he cannot make the expiration; inspiration, when again made, was found easy enough, but attended by a slight crowing sound, expiration again impossible. The face was livid, with blueness of the mouth. The lungs are fearfully distended from frequent inspiration without any corresponding exit of air, and this finally results in more or less complete asphyxia, with or without convulsions, during which the spasm relaxes and free respiration takes place. The attacks come on after excitement, during sleep, and they are most common from midnight till seven A. M.

Dr. Dunham, who introduced Chlorine to our notice, gives us the following instructive cases:

"June 24th. A female infant, seven months old, well developed and large, the fourth child of healthy parents, was brought to me with the following history: Having been previously in perfect health, she was seized three weeks ago with a spasmodic affection of the respiratory organs. Suddenly, and without any warning, she would make a long inspiration, with a slight crowing noise; an attempt to exhale would be made, but without success; another crowing inspiration followed by a forcible but ineffectual effort to exhale, and this would be repeated until the child became blue around the mouth, and sank into partial unconsciousness, when free respiration would take place, and the child would generally sink into a deep sleep. Frequently towards the close of an attack, convulsive movements of the extremities would be noticed, and once general spasm occurred. At first these attacks came on only after some excitement, or on the child being startled. They frequently occurred during sleep, arousing the child suddenly, and they were most frequent from midnight to 7 A. M. Within the week before I saw her, they had become very frequent—as many as 30 to 40 occurring during the 24 hours. The child had begun to emaciate rapidly, had lost appetite, strength and playfulness, the face was pale and bloated, and the eyes had a dull and glassy expression. The child had been under most skillful homœopathic treatment since the commencement of the attacks, and as she failed to improve, change of air was recommended, and she was brought to Newburgh. The climate failing to benefit her, the child was placed under my care. The case seemed all the more serious from the fact that, last year, the parents had lost an older child, a boy, with the same affection. In the fourth week of the disease, of which the course had been in every respect similar to that of the infant above narrated, convulsions supervened, and the child died at the end of the sixth week. This child was under enlightened allopathic care. It may be interesting to note that the autopsy revealed no malformation, and no organic lesion; simply emaciation and atrophy.

"On careful examination of my little patient, I could discover nothing abnormal in the condition of the heart or lungs, and no sign of disease that was not fairly attributed to the frequent recurrence of these spasms, with the venous congestion consequent upon them, It was evidently a case of Spasmus Glottidis (asthma thymicum, asthma millari, asthma laryngeum infantum, laryngismus stridulus), and had. advanced almost to the second or convulsive stage in which the prognosis is decidedly unfavorable.

"The remedy which is recommended before all others for this disease, in our hand-books and repertories, is Sambucus. The symptoms on which this recommendation is based are the following: 'Slumber with half-open eyes and mouth; on awakening from it he could not draw a breath, and was compelled to sit up, whereupon respiration was very hurried with wheezing in the chest, as if he should suffocate; he lashed about with his hands; the head and face were bloated and bluish; he was hot without thirst; weeping at the approach of a paroxysm; all this without cough, and especially at night from twelve to four o'clock.' On comparing this picture with the case under consideration, we find correspondences in the general characters of the affection. The spasmodic embarrassment of respiration, the absence of fever and of cough, the occurrence of the paroxysms *suddenly*, chiefly at night, and on awaking show a general appropriateness of Sambucus to spasm of the larynx and bronchial tubes. But we seek in vain for the unequal disturbance of the *inspiratory* and the *expiratory* act, which are the *individual* and therefore the *characteristic* peculiarity of the case under consideration. And failing to find this, we should, as a matter of course, expect that Sambucus would fail to cure, or in any way to affect the case. And this had been the fact. So, too, of Lachesis and several other remedies which, as well as Sambucus, had already been tried before the case came under my care. In this very peculiarity, which was characteristic of the case, the similarity of Chlorine was most striking. And it was with the utmost confidence of a happy result that I determined, after a careful examination of the case, to administer Chlorine. I accordingly prepared a saturated solution of Chlorine gas in water of 60° Fahrenheit, and made from this the first centesimal dilution in which the odor of the Chlorine could be faintly perceived.

"Of this, I ordered twenty drops to be dissolved in four tablespoonfuls of water, and a teaspoonful to be given to the child every three hours (a porcelain spoon was used). I also directed a few drops to be placed in the child's mouth at the beginning of each paroxysm if this should be possible.

"The first dose was given at four P. M., June 24th. During the preceding twenty-four hours the child had had forty paroxysms. During the succeeding twenty-four hours, there occurred but four paroxysms; only one of which began with any severity, and this one was instantly arrested midway by a few drops of the solution placed upon the child's tongue. During the night of the 26th, not a single paroxysm.

Improvement in the general condition of the patient now became apparent, appetite and playfulness returned, the bloated aspect of the face and the dulness of the eye disappeared. On the 27th, the paroxysms increased in number and severity. On examining the solution I found that it had changed in character, and no longer contained Chlorine. A fresh solution was prepared, and henceforward it was prepared fresh every second day. From this time, July 1st, the remedy was continued; a dose every four hours—when the spasms having wholly ceased, and the child appearing well, it was finally discontinued. On the 2d of July a slight spasm occurred, and the child appeared feverish and excited—with greenish diarrhœa. I found a lower incisor pressing strongly upon the gum, which was hot and swollen, and which I forthwith lanced. In two hours the child had lost every trace of illness. Since that date she has continued in good health, with the exception of some trifling disorder attendant upon dentition. There has been no sign of a recurrence of the spasm of the glottis."

"Last month I was called to a child two years and a half old, which had just been brought home from the country, and was supposed to be at the point of death. Five weeks before, it had sickened with scarlatina, which, according to the physicians in attendance, had become complicated by diphtheria, and this by inflammation of the right lung and deposit therein. An abscess had formed and discharged externally on the neck, leaving an ulcer about two inches long and one and a half broad, which exposed the cervical muscles and showed no disposition to heal; copious and very offensive discharge from both ears; the throat seemed to be full of a thick, yellow matter, very offensive, which the child would occasionally eject, but seemed, for the most, to be unable to move either up or down. Any attempt to examine the throat, or, on the part of the child, to open the mouth to take food or drink, or any attempt to cough, produced a fearful spasm of the glottis, which seemed to *admit* the air well enough, but to prevent its *exit*, and which lasted until the child became livid and sank exhausted, and this constituted, in the opinion of my predecessors, the insuperable obstacle to the child's recovery. The spasm prevented its tasting food. No food had been taken for a week, and very little during the entire illness. The child was now very feeble and greatly emaciated. Its death had been hourly looked for by the doctors.

"I prepared immediately some *Chlorine water*, diluted, until the gas was just perceptible, and gave it to the child. He took a mouthful and began to choke with the spasm; I immediately placed near his face a handkerchief wet with strong Chlorine water, so that he might inhale the gas. The spasm ceased instantly and he swallowed. I left orders for a similar procedure whenever, from any cause, whether coughing or swallowing, the spasm should be induced. It never failed to arrest the spasmodic action and enable the child to swallow, or to eject the matter from the throat. A number of days elapsed before the child could make an articulate sound, or any sound. The doctors had thought the diphtheria had induced paralysis of some of the pharyngeal muscles, and perhaps others, and hence the spasm in associate and neighboring muscles; and it may be so. They regarded the spasm as an insuperable barrier to recovery. It was evident to every attendant that the Chlorine relaxed the spasm and enabled the child to swallow. His subsequent improvement was uniform and rapid under Carbo vegetabilis 200."

Cuprum is, according to Bæhr, particularly appropriate if, during the local spasm, general convulsions have supervened and the child has become very much prostrated, and Hughes remarks that whereas Belladonna should be given where there is arterial excitement and cerebral congestion, "Cuprum should be given where these symptoms are absent"—that is, when the morbid state is a pure neurosis. Farrington thinks Cuprum well adapted to cases which have advanced to the convulsive stage, and Jahr advises that if the spasm of the glottis sets in in company with other spasmodic symptoms to give "above all Cuprum," and I have certainly never seen any good effects from this remedy save under these circumstances. Duncan, too, tersely says " Cuprum is the remedy." Cuprum is indicated by short, panting, whistling breathing, with gurgling down the œsophagus, and on attempting to take a deep breath there is a dyspnœa, with stridulous inspiration. This local, morbid state is accompanied by general convulsions, the body being stiff with spasmodic twitching and clenched thumbs. The face and lips are both alike blue, and the face is sometimes covered with cold sweat. The paroxysms come on suddenly and cease suddenly, after fright of mother or child. Searle calls attention to cold perspiration at night as being a kind of key note, and Bæhr says that among the symptoms indicating the remedy one is particularly noticeable—vomiting after the attack.

"A delicate girl, nine months old, had for several days suffered with a cough, spasmodic and more violent during the night, peevishness, no fever, quick, difficult breathing, drawing in of the muscles of the right and left hypochondriac regions during inspiration, percussion normal, rattling of mucus far down, little appetite, tongue with whitish coating, daily, one or two thin, sometimes watery, sometimes greenish stools. Ipec. 9 every two hours. While asleep the child suddenly began to breathe more quickly, and with greater difficulty; grew restless and tossed about in bed; face bluish; eyes wide open; larynx drawn upward; she braced herself against the bed with her hands; perceptible cramps in the respiratory muscles; predominant abdominal respiration; the cough, which was very exhausting, was attended by a very peculiar hollow, somewhat hoarse sound; at times, also, metallic-sounding, piping, short coughs; hands cold; cold sweat on forehead; spasmodic, small, very frequent pulse. The attack lasted five to six minutes; afterward the child sank back exhausted, coughed a few times loosely and easily, and fell into a stupefied sleep. She had five or six of these attacks for several consecutive nights, but of longer duration. Ipec. every two hours. The next day only one attack, which lasted only three to four minutes. During the day great debility, little appetite; cough easy and loose, and even none at all for four or five hours at a time; respiration normal; two somewhat slimy, but otherwise healthy stools. The next night two rather lighter attacks, but next day still greater debility. *Cuprum* 9 in *Sacch. lact.*, one powder; if necessary another during the night. At midnight a very light attack, lasting only two to three minutes. The next day general health and appetite better. One dose *Cuprum*. No more attacks and soon restored to perfect health. (Dr. Hirsch.)

"A very delicate child, about one year old, had, since six nights, very violent attacks of spasm of the glottis without any coughs, either during the attacks or at other times; they lasted five to ten minutes. Cuprum 9, three doses, one every evening, relieved the patient entirely." (Dr. Hirsch.)

I have never seen any good results from Cuprum when given below the 12th centesimal dilution, and have generally given the Hahnemannian 30th with almost unvarying success, but I must add that, in my experience, cases in which Cuprum is indicated are not very common.

Hughes says that in the paralytic variety *Ignatia* seems to be the remedy most homœopathic to the paroxysm, but Bæhr, after stating that this remedy is very much praised if the children suddenly lose their breath, which may be the lowest degree of spasm of the glottis, adds that "whether it will prove a proper remedy for spasm of the glottis has not yet been verified." Personally, I have not found it useful in idiopathic cases, though it has helped when the spasm was a mere incident in other diseases, say catarrhal croup or whooping cough. Of the symptoms one of the most characteristic is the difficulty of inspiration while expiration is easy, and this difficulty is suddenly experienced at midnight. All kinds of respiration alternate during sleep, short and slow, deep and light, intermitting and snoring. Also a sudden (not tickling) interruption of breathing in the upper part of the trachea, which irresistibly provokes a short, forcible cough in the evening. I have always used the 12th centesimal dilution, though possibly the 6th might have been still more effective.

Lachesis has been successfully used, though Jahr says it has never afforded him much help in this disease, while Drysdale says that it is one of the two medicines which he has found, on the whole, most useful—arsenicum being the other. Personally, I am of opinion that, like Ignatia, this remedy is not indicated in idiopathic spasm of the glottis, but that it is often indicated when a partial, imperfectly-developed form of this morbid state supervenes on inflammatory affections of the throat.

Lachesis is indicated when the paroxysms occur during sleep; the child, as it were, sleeps into an attack, and is roused gasping for breath. Or the paroxysms may recur after each nap. There is great sensitiveness of the larynx and trachea to the touch; sense of constriction of the larynx, attended with dryness of the whole throat and mouth. Dr. S. C. Knickerbocker reports an excellent case in which the attacks were becoming more frequent and more severe, which was radically cured with two prescriptions of Lachesis 200. The attacks consisted of a sense of constriction of the larynx, attended with dryness of the whole throat and mouth—the attacks invariably occurred after sleeping—the "key-note" of Lachesis.

"Plumbum is very closely related to Cuprum in every respect, except that the general strength is still more reduced. The symptoms of a spasmodic closing of the rima glottidis are more distinctly marked in the pathogenesis of this drug than in that of any other. We are amazed that Plumbum should not yet have been recommended for this disease, which, however, can only be cured by remedies which exert a deeply-penetrating, long-lasting influence over the whole organism." (Bæhr.) Kane, too, recommends it the last of the few but well-chosen remedies with which he combats spasm of the glottis, but Searle, after remarking that it has the mucous râle with sudden difficulty of breathing and asphyxia, adds, "but I do not know that it has ever been tried as a remedy for this disease." Here, too, I have had no experience, for I have never met with a case in which it seemed to be indicated.

Dr. Alphonse Teste considers that Corallia rubra and Opium are "heroic agents against this disease," and as to the dose, he adds, "I prescribe Corallia at the thirtieth and Opium at the third dilution, a teaspoonful every two hours during the period of invasion; every ten minutes during the exacerbations, and at intervals gradually increased when these are passed. The last-mentioned remedy is to be given every six hours, for a day or two after the resolution of the last attack." As spasm of the glottis frequently accompanies rachitis, Dr. Richard Hughes remarks that "the Corallia rubra so lauded by Teste, in its treatment may, from its calcareous nature, be suitable to these diathetic conditions as well as to the laryngeal spasms." Searle thinks that the Corallia rubra may be serviceable in cases which it is difficult to distinguish from whooping cough, and Farrington recommends Opium "especially after a fright," but I have never seen any benefit from the use of Opium in this disease.

The above are the leading remedies, but occupying a secondary rank are Zincum, Bromine, Nux vomica, Pulsatilla, Veratrum album, Laurocerasus, Spongia and Sulphur.

Dr. Pemerl, of Munich, points out that "when the spread of the spasms of the glottis to other respiratory muscles, to the tongue, to the upper and lower extremities, announces the transit to general convulsions, Moschus does not suffice any more, and we prefer the first trituration of *Zinc-oxyd.*, either alone or in alternation with Moschus. Bromine has spasmodic closure of the glottis and constriction in the larynx, with a wheezing and rattling in the larynx; gasping and snuffling for breath; cannot inspire deep enough; the head and face are hot. *Nux vomica* is recommended by Kane and Duncan for reflected irritation from derangement of the digestive organs. Pulsatilla is said by Laurie to be often found successful in cases in which Moschus appears capable only of effecting a limited degree of improvement, but the present writer has never seen any good effect from it, and it is doubtful whether it has ever been successfully used in unquestionable cases of spasm of the glottis. *Veratrum album* is recommended by Jahr if spasm o the glottis sets in in company with other spasmodic symptoms; Bæhr advises it, together with Arsenicum album, if the disease attacks feeble children with marked symptoms of anæmia, and the writer has used it successfully, in the 12th centesimal dilution, when the patient was already cold and pale, with contracted pupils and protruded eyes. *Laurocerasus* when the child is pale and blue, with spasmodic constriction of the throat and congestion of the chest; Farrington recommends it when the head is affected, which is rarely the case. *Spongia* is indicated when the child starts from sleep with constriction of the larynx, whistling respiration, the patient breathes with the head bent backwards; this remedy may be used against the constitutional disease as well as the laryngeal spasm. *Chamomilla* is recommended for a sensation of oppression and slight constriction in the region of the larynx; constriction of the larynx with dyspnœa; hot sweat on the head and face, especially during sleep. The child becomes stiff and bends backwards, kicks with his feet when carried, screams and throws everything off; staring eyes, the child reaches and grasps for something, draws the mouth back and forth. The patient is peevish and irritable; cries for things and pushes them away when given to him; worse from anger or other violent emotions, from dentition and from exposure to cold winds. This remedy is not suited to well-developed cases, though I have seen it do good in such cases as the above. Of *Sulphur*, Bæhr says that it " may deserve attention, although we shall take the liberty of doubting the homœopathicity of its asthmatic symptoms to spasm of the glottis until the fact has been corroborated by experience," but though I have never seen a case in which Sulphur would help during the acute attack, I have for many years given it in the 30th dilution to prevent a recurrence of the dreaded disease, thus following an invaluable hint of good old Jahr's. "In more than one case, however, I have radically removed a disposition to the return of the spasm by means of a dose of Sulphur."

The remedy indicated by the *ensemble* of the symptoms should be given at the time of the attack as well as during the interval, but it is evident that, as the paroxysm lasts but a very short time, that it is not always possible to give the medicine. The same remark applies to very many of the external applications recommended by various authors, for, in the vast majority of cases, the little patient is out of the paroxysm before the machinery is in motion, and in the matter of treatment, the writer relies very much on what is done *during the interval*.

All external causes which may have the effect of irritating or exciting the nervous system should be carefully avoided, and moral causes are just as important as physical. There should be no sudden surprises, either playful or otherwise. The temper should be irritated as little as may be. All muscular effort should be carefully avoided, and Copland gives the excellent advice to avoid straining at stool.

During the attack the patient should be placed in an upright attitude, so as to place the larynx in the easiest position possible, and all tight clothing should be promptly removed—or rather, tight clothing should never be put on children subject to spasm of the glottis. Vogel thus describes a simple operation much in vogue in Germany, which I have used with marked benefit: "In the instances where I happened to be present at the paroxysms, I introduced the index-finger into the mouth, carried it to the posterior pharyngeal wall, elevated the spiglottis, and then touched the chordæ vocales, by which marked acts of choking were at once induced, and then the well-known whistling inspiration followed. Lay people, of course, are unable to execute these manœuvres, and I therefore content myself by showing them how retching may invariably be induced by pressure upon the root of the tongue. The shock produced by inducing this act of retching is the only harmless remedy which will cut short the paroxysm." Dr. Morell Mackenzie's advice to "slap the patient on the back" should be carefully shunned. The same writer advises the dashing of cold water on the face, though Vogel says that he has seen no decided effects from cold affusions as well as from the forcible to-and-fro swinging in the air, so much in vogue with nurses. Steffen advises the full warm bath of from 90° to 95° Fahr. combined with cold affusions over the head and neck, if the cyanosis assumes a high grade, consciousness is lost and general convulsive attacks set in, but the one objection to these procedures is the time involved in carrying them out. Romberg gives us the excellent advice to warm the præcordia with hot napkins during the attack.

Should Chloroform be used during the paroxysm? Sir James Simpson proposed its use in this disease, and Dr. Charles West and other excellent English practitioners say that they have always secured prompt results without any ill effect whatever; Dr. Duncan of Chicago, an excellent homœopathic authority, endorses this recommendation. Morell Mackenzie says that "the inhalation of Chloroform is a very valuable remedy, but of course, it must be used with great care." Steffen, too, advises it to be used with great caution, and Dr. G. B. Wood advises the practitioner to bear in mind the hazardous character of this remedy. Vogel remarks that "it is too dangerous an agent to be left to the use of the lay attendant," and of course the physician could rarely be present to administer it during the paroxysms, and for this reason as well as for that given by Romberg, I am disinclined to use this agent. Romberg says, "it has been proposed to give Chloroform, but its effects upon the brain under such circumstances would probably render it unsafe."

Should tracheotomy be employed as a last resort? Marshall Hall, together with Wunderlich and other excellent German authorities, recommend it in the last emergency, when suffocation is taking place, and one of the latest and best of the English writers strongly advises it. "If the child appears to be sinking from the apnœa, the trachea must, of course, be opened, and artificial respiration resorted to. Indeed, this should even be adopted by the practitioner, should he arrive shortly after the apparent extinction of life." But Steffen, a still higher authority in this particular disease, remarks that, "aside from the fact that, unless the patient happens to be in an hospital, this operation cannot always be performed quickly enough, I have never yet learned of any favorable result that has followed in spasm of the glottis," while Vogel says that "tracheotomy, which has been suggested as a *dernier ressort*, with which to save the life of the child, can never be performed, on account of want of time." Partial as are the French to tracheotomy, they have never, so far as I know, recommended it in this disease, and to me the reasoning of Romberg is perfectly conclusive, "nobody would attempt tracheotomy at the beginning of the attack, and if postponed too long no benefit can be expected from it."

Vogel, Steffen, and all the great German writers on this disease, condemn the lancing of the gums, and Romberg's dictum may be taken as a fair specimen of their reasoning: "Scarification of the gums, in England considered a panacea, has not met with much countenance in Germany, as the excitement produced by the operation in the child outweighs the possible advantages of the operation." On the other hand, almost all the British and American writers approve of the operation, and we have seen that it was performed with success by Dr. Carroll Dunham, one of the strictest of Hahnemannians. I have never performed the operation, simply because I have not yet met with a case in which it was indicated.

During the intervals of the paroxysms, and still more during the paroxysm itself, there should be a good, long interval between the feeding hours of the child, and during the continuance of the disease the infant *must not be weaned*, and I have heard of more than one death from ignorance or neglect of this simple and almost self-evident rule. Vogel advises that the child be kept as long as possible at the mother's breast, at least until it has cut the first six incisors. "If the fit comes on during sucking, either from the leather teat of the bottle or whilst the child is at the breast, it must be fed as Flesch insists, with a very small teaspoon, no matter how difficult at first it may be to get nourishment taken in this way." (Morell Mackenzie.) No food should be given soon after a paroxysm, for a second paroxysm may result from the mere act of swallowing, especially if particles of food enter the larynx. To children five or six months old I give beef tea, not essence of beef, but a weak preparation made by boiling finely minced beef with a considerable quantity of water, straining it through fine muslin or blotting paper, and I have seen good results from the addition of a soft-boiled egg to the diet. If the malady is complicated with rachitis, no farinaceous food whatever should be given; such patients should be fed on meat and milk till they are at least seven years of age. Steffen advises the administration of Hungarian wine, or good, French red wine.

A child subject to spasm of the glottis should never be disturbed during sleep, as, in many cases, the excitement of awakening brings on a paroxysm. Steffen thinks that children should not be kept warm in bed, and, in opposition to Elsässer, he asserts that lying down is no sort of disadvantage to children with craniotabes.

I have found the cold sponging of the chest advocated by Dr. Richard Hughes more effective than the daily luke-warm baths of Steffen, and as the child gets older I recommend free sponging of the entire body with cold water every morning.

To patients living in a city or large town, great benefit accrues from removal to the fresh, pure air of the country, and if the case is at all severe the patient should be at once removed. Still, as Vogel remarks, residence in the country by no means supplies a positive guarantee against the appearance of the spasms, and some of my worst cases occurred in children who had been all their lives in the country. Wherever the patient is the rooms should be carefully and systematically ventilated, and Robertson recommends the free exposure of the infant out of doors for many hours daily, to a dry, cold atmosphere, and if the air be dry the colder the better. Dr. Marshall Hall says that the curative influence of change of air, and especially of the sea-breezes, is not less marked in this affection than in whooping cough.

Aphorisms.

1. Spasm of the glottis is a constriction of the muscles which narrow the glottis, accompanied by crowing inspirations, commencing suddenly, lasting a very short time, and ceasing suddenly.

2. General convulsions and "carpo-pedal" spasms mark the advanced stage of the disease.

3. Spasm of the glottis may depend upon an irritated disease of the brain, and also upon the scrofulous and rachitic constitutions—though its connection with rachitis is less clear than its connection with scrofula.

4. The disease is most prevalent in northern countries and during the Winter season, and it is essentially a disease of childhood, though adults are not exempt.

5. Two-thirds of the sufferers from spasm of the glottis are boys.

6. The disease is not really hereditary, though several children in the same family may suffer from it.

7. Rachitis and spasm of the glottis often co-exist, but the first is not neccessarily the cause of the second, though rachitis of the thoracic bones may lead to the disease under consideration by inducing a deep-seated disturbance of nutrition and an increased irritability of the nervous system.

8. Weaning favors the development of spasm of the glottis, and the irritation of teething may cause an outbreak of the disease.

9. As the child grows older the predisposition to the disease declines, and this depends on the increased size of the larynx and on the decreased irritability of the nervous system.

10. The prognosis is more favorable in girls than in boys, and the older the patient the better the prospect of a successful issue.

11. The leading homœopathic remedies are Sambucus, Aconite, Sanguinaria, Arsenicum album, Belladonna, Gelsemium, Iodium, Chlorine, Cuprum and Ignatia amara.

12. Chloroform must never be used by lay hands, and it is positively dangerous in cases depending on cerebral irritation.

13. The weight of evidence is against tracheotomy as a remedy *in extremis*.

14. The patient must not be weaned during the continuance of the disease.

CHAPTER V.

ACUTE CATARRHAL LARYNGITIS.

The laryngeal diseases of young children are always very serious, as from the small size and delicacy of the organ in infancy, a comparatively slight inflammation greatly diminishes its calibre. Again, the organ is absolutely essential to life, and but a slight disturbance of its healthy function is enough to endanger the very existence of the child, especially in those not rare cases in which there is a hereditary proclivity to laryngeal diseases. Laryngitis is somewhat frequent during infancy and childhood, and I cannot agree with M. Bouchut who considers it "a disease of slight importance," adding that its termination is "always favorable." On the contrary, it is frequently a serious disease, coming on suddenly, attacking violently, and requiring skillful treatment from the physician and careful management from the nurse.

The disease may be defined to be a catarrhal inflammation of the mucous membrane of the larynx, sometimes involving the submucous areolar tissue, giving rise to hoarseness or aphonia, stridulous and difficult breathing, cough and pain in the larynx—especially near the *pomum Adami*. Dysphagia is present in very severe cases, and fever is an almost invariable accompaniment. If the inflammation is confined to the middle and lower parts of the larynx the cough will not be croupous, but if the epiglottis and rima glottidis are affected the cough will be decidedly croupous, and hence, many writers and practitioners speak of the disease as being a variety of croup. In reality, as Dr. W. V. Drury has well pointed out, every one, and especially every mother, should know that there are five or six different diseases with a croupal cough—acute laryngitis, spasmodic laryngitis, membranous laryngitis, dipthertic laryngitis.

Acute catarrhal laryngitis is often the result of repeated congestion of the larynx, and it may follow any irritation of the mucous membrane, which irritation results in engorgement of the blood vessels, swelling and succulence of the mucous membrane, with copious generation of cells and an abnormal amount of mucus secretion. There are two varieties of acute laryngitis—acute catarrhal laryngitis and acute œdematous laryngitis—the first affecting the mucous membrane only, the second affecting the sub-mucous areolar tissue. The first mentioned also occurs in connection with some of the infectious fevers—measles, scarletena and variola —and, to some extent, it is present in most cases of bronchitis and even in pneumonia.

Hippocrates makes some mention of a disease which was most likely laryngitis, but we find no other mention of it till the eighteenth century, when it appears to have been recognized by Drs. Mead and Millar, though the latter obscured his picture of the disease by confounding it with the spasm of the glottis to which his name has been attached. Later it was described by Dr. Hume in his *Principia*, and in 1809 Dr. Baillie gave a very full account of it. The first dissection of the disease appears to have been made by Mr. Mayd in 1789, and forty years later Guersant first gave a clear account of its pathology.

Dr. Ellis, of Auckland, New Zealand, says that this " is a disease far commoner in adults than in children ; still it does occur in children ; " but on our continent, at least, it is a very common disease amongst children, especially if the variety affecting the rima glottidis is taken into consideration. Acute catarrhal laryngitis occurs more frequently in children under five years of age than in those over that age, and Duncan remarks that of sixty-two well marked cases met with, in which the age was noted, fifty occurred in children under, and only twelve in those over that age; most of the fifty cases were under two years of age. It rarely attacks children at the breast, though it is not rare in children six or seven months old. It prevails in the fall, spring and winter months—especially in March and April—though severe cases are sometimes met with in summer. For obvious reasons, it is more frequently met with in boys than in girls.

Relaxing habits and confinement indoors undoubtedly predispose to this disease, and children resident in towns are more likely to be attacked than those living in the country. Long continued and violent crying often causes the disease, and it may follow the inhalation of dust or contaminated air. But the most common cause is "taking cold," and it often follows sitting in a draft of cold air, or permitting the child's feet to remain wet and cold. Quite often a severe coryza extends downwards to the larynx, and, but more rarely, the morbid process extends upward from the bronchial tubes. After a child has once suffered from acute catarrhal laryngitis other attacks are almost inevitable.

The disease usually commences as a common cold; there is chilliness followed by fever, with slight sore throat. Sneezing is often present, with slight hoarseness, all of which symptoms point to a somewhat mild catarrhal affection which may suddenly become a serious malady with symptoms of the gravest. Or, in the midst of excellent health, it may appear suddenly in the night, but, as a general rule, the above-mentioned prodomata are present. These symptoms persist for two or three days, when suddenly the voice becomes hoarse or disappears altogether, showing that the larynx is involved, for in that organ, and there only, is the voice formed. In the other class of cases, the attack is sudden, and the larynx is affected from the first. The child goes to bed apparently almost well, or with but a slight sneezing or coughing, when, after two or three hours of sleep, he wakes up very ill indeed. There is a hoarse and barking cough, ringing and croup-like, and accompanied by expectoration; it is paroxysmal and worse in the evening and during the night. The cough is wholly laryngeal, and sometimes a little tough mucus is raised with relief of the local symptoms. If the patient is able to describe his feelings, complaint is made of a dull, aching pain in the upper and front part of the throat, with a marked feeling of constriction, which prevents the patient from using the voice, even before it disappears; there is difficulty of swallowing, as much from the pressure of the inflamed larynx as from actual pharyngitis; the larynx is felt to be enlarged, hot and tender on pressure, and the difficulty of breathing which is present from the commencement becomes aggravated. The little patient often puts the hands to the throat, and at times there are spasms of the muscles of the glottis, from which the disease gets the name of catarrhal croup. The child is very thirsty, restless and uneasy; the skin is dry and hot; the pulse full, and from 100 to 120 in the minute. This fever, with rapid pulse, hot skin and scanty urine, assumes the asthenic type, from carbonic acid poisoning, if the disease is unchecked. When croup-like cough, with markedly croupous breathing, hoarseness and fits of choking are present, it is almost impossible to distinguish the disease from pseudo-membranous croup. But after the symptoms have lasted for an hour or two, the breathing becomes normal or nearly so, the hoarseness almost disappears, moist râles appear in the chest, general perspiration breaks out, and the child falls into a sound sleep, usually accompanied by loud snoring. The most marked characteristic of the disease, then, is the paroxysmal appearance of the stenosis of the larynx in the night, alternating with the symptoms of catarrh of the larynx during the day. This depends upon the fact that during the night the copiously-secreted mucus settles in the very narrow glottis, in fact almost closing the rima glottidis, at the same time adhering to the vocal cords. This, of course, gives rise to a rapidly-increasing impediment to respiration, till finally, after coughing and crying, and, it may be, vomiting, the mucus is removed for the time. Next morning the laryngeal stenosis has wholly disappeared, and is replaced by the catarrhal laryngitis with slight hoarseness, and, though in some instances no second attack comes on, it frequently returns the next night.

If the disease should prove uncontrollable, the breathing becomes still more obstructed, and inspiration requires an unusual effort, and is hissing and whistling. The cough becomes still more distinctive in its character. " It is brassy in its tone, terminates in a hissing noise, and begins similarly by a hissing inspiration in a muffled manner, because the lips of the glottis being thickened, irregular and rough, cannot be sufficiently closed to begin a sharp sound." (Hyde Salter.) " As the aperture of the glottis becomes narrower a terrible picture of distress presents itself, for strangulation seems to be imminent, and the patient tosses himself anxiously about, gasping for breath ; the face is pale and livid, the eyes start from their sockets, and the poor sufferer asks for fresh air, walks about, and goes to the window for it, and finally delirium and coma close the scene ; in fact, to use the expression of an able observer, he dies strangled." (Gibb.) Hoarseness remains for a number of days, and in the morning the cough is so violent and prolonged that it is sometimes difficult to convince the parents that the child is safe. I have seen cases in which, after repeated attacks of catarrhal laryngitis, an attack of true croup came on, under which the child succumbed, and this is quite likely to occur if such a patient is attacked with measles. In other cases again, chronic laryngitis resulted from repeated attacks of the acute disease. If the inflammation is very severe, œdema of the glottis may supervene, and the possibility of this untoward event must always be kept in view.

There is a great difference of opinion as to the duration of this disease. Ellis thinks that it ought to disappear in from four to six days ; Vogel gives from three to eight days as the average duration ; Von Niemeyer says that it ought not to last more than a week ; Behr says that it lasts, at most, nine days ;·J. Lewis Smith thinks that it disappears in from one to two weeks ; and von Riemssen gives for mild cases five to seven days, moderately severe, eight to fourteen days, while the most severe, according to this writer, run from two to three weeks or longer. Other excellent observers take a more gloomy view, for DaCosta writes as follows: "The disease in its graver form runs a very rapid course. If, in a few days after its commencement, no improvement show itself, life does not last long. Sometimes death takes place on the first day of the attack. It rarely waits for the sixth." Morell Mackenzie is only a little less gloomy: "The acute stage seldom lasts more than three or four days, and I have seen a case terminate fatally in twenty-four hours. Death has been known to occur in seven hours. It is rare for the symptoms to remain serious after the fifth day, unless a kind of chronic œdema sets in." My own experience is that the duration of an acute case is from four days to a week, and decided danger is not far off if the disease is permitted to run on in a severe form much longer than a week, and this danger may take the form of true croup. If the disease is uncured, a chronic inflammation, or rather congestion, of the larynx remains, which is somewhat difficult of cure, though it must be admitted that spontaneous resolution, or resolution as a result of therapeutic interference, is far more common. When death takes place it is usually the combined result of spasm of the glottis and œdematous swelling, and the fatal event is often preceded by carbonic acid poisoning with its accompanying delirium.

The thermometer does not usually show a marked rise in temperature during the day, but at night the increase is very marked, and the 99° or 99.5° of the day advances to 102° or even 103°. The appearance of laryngeal spasms, not being due to inflammation, does not cause much alteration in temperature.

On examining the larynx it is found to be of a bright, cherry red color, which also extends to the tonsils and soft palate. The mucous membrane of these regions is dry and swollen, and the papillæ are more prominent than in health. The epiglottis is of the same bright, red hue, and when felt by passing the forefinger down the throat it gives the sensation of a round body of the size and consistence of a ripe cherry. As a general thing the whole of the pharyngeal mucous membrane is of the same bright, red color, and it is greatly congested and swollen. The vocal cords are of the same color, though they often have patches of a darker shade, and, later in the disease, a thick, adherent mucus covers all the mucous membrane, giving it a grayish tint. The dyspnœa and hissing respiration are caused by the swelling of the vocal cords, and not by the narrowing of the glottis as usually supposed. Gerhardt points out that the hoarseness is often the result of a partial paresis of the thyro aretenoid muscles, and this often precedes the congestion of the laryngeal mucous membrane, so that the old observation that hoarseness in a child is of more serious import than hoarseness in an adult may be looked upon as an established fact.

The *post-mortem* appearances are confirmatory of those observed during life, though, as Felix Niemeyer remarks, the mucous membrane of the cadaver does not always reveal a degree of redness and vascular engorgement such as the violence of the symptoms during life would lead us to expect, and such as could then be demonstrated by laryngoscopic observation. This, according to the same acute observer, is due to the richness of the laryngeal mucous membrane in elastic fibres, which, remaining distended by the blood contained in the vessels during life, after death contract, and expel the contents of the capillaries. The disease is a simple inflam. mation of the mucous membrane of the larynx, sometimes involving the sub-mucous areolar tissue, and accordingly there is reddening and swelling of the laryngeal mucous membrane, with a coating of mucus. It is rare that the larynx only is affected, for in almost every case the morbid process extends to the pharynx and trachea, producing very similar *post-mortem* appearances. At times the mucous membrane of the larynx is abraded from the removal of the ciliated epithelial cells, but true ulceration is rarely present ; if ulcers are present they are most likely syphilitic or scrofulous in their nature, and were present before the invasion of the acute disease.

The diagnosis of this disease is easy, though sometimes it is difficult to distinguish it from pseudo-membranous croup, for an apparently simple inflammation may really be the early stage of the plastic form of the disease. In catarrhal laryngitis the pharynx and tonsils are simply reddened; in pseudo-membranous croup an examination of the fauces reveals patches or a continuous coating of pearly-white exudation on the soft palate, half arches, tonsils and pharynx. In catarrhal laryngitis the lymphatic glands of the neck are normal; in pseudo-membranous croup they are often swollen. In catarrhal laryngitis the fever remits during the day, but in pseudo-membranous croup the remission, if any, is very slight. Indeed, it may be said that the entire morbid process of catarrhal laryngitis is remittent in its character, for, though the voice remains hoarse with a somewhat clangorous cough, the local affection and the fever are so slight that many children amuse themselves as if nothing were the matter. All this is completely reversed in pseudo-membranous croup, and both the local affection and its accompanying fever are much more pronounced. Catarrhal laryngitis is at first unaccompanied by expectoration; and, as amendment sets in, a slight expectoration of ordinary mucus appears; while the only sure diagnostic sign of pseudo-membranous croup is the expectoration of fragments or tubes of false membrane.

Catarrhal laryngitis may be confounded with spasm of the glottis, but attention to the following points will at once clear up the difficulty: Catarrhal laryngitis comes on gradually, while spasm of the glottis is marked by a sudden accession. Catarrhal laryngitis has hoarseness during the attack, which persists during the interval; spasm of the glottis has no hoarseness at any time. Catarrhal laryngitis has a croupy cough, worse at night, better during the day; spasm of the glottis has no cough whatever. Catarrhal laryngitis is unaccompanied by constitutional fever; spasm of the glottis very rarely has fever. Acute catarrhal laryngitis may be distinguished from whooping cough by the fact that the latter disease has no hoarseness, no inflammation, no fever and no thickening of the mucous membrane of the fauces.

A great variety of opinions exist as to the prognosis of this disease, and that, too, amongst writers who are clearly describing the malady. Ellis thinks that the prognosis is "very unfavorable," and Dr. Morell Mackenzie says that "in early life, that is before the development of the larynx has taken place at puberty, the disease is always attended with great danger." According to the same author, it is more fatal in children than in adults, "more than four-fifths of the mortality occurring before the tenth year." On the other hand, Felix von Niemeyer says that "a fatal termination, uncomplicated by any other cause of death, is one of the greatest of rarities," an opinion which is echoed by Bæhr; von Riemssen, too, affirms that "a fatal result is extremely rare," while Steiner thinks that "the issue of acute laryngeal catarrh in recovery is the rule almost without exception." My own experience is that the prognosis is generally favorable. I have never known a fatal case under homœopathic treatment, and I could scarcely imagine such a patient dying under the care of a well-read physician of our school. Favorable signs are a diminution of dyspnœa, freer expectoration, and less difficulty in swallowing. The supervention of œdema of the larynx, fortunately rare, would greatly darken the prognosis. Very much depends upon the stage at which the physician is called in; should coma be present, there is little hope. Previous allopathic treatment would diminish the chance of recovery, especially if emetics and mercurials have been used. Still, it must always be borne in mind that, even in healthy children, a simple catarrhal laryngitis may be converted into a true pseudo-membranous croup, and this possibility should always be present in the physician's mind. This is quite likely to take place in children suffering from measles or small-pox.

In the treatment of this disease, Vogel's warning should

ever be present in the mind of the physician; "Pseudo-croup should never be regarded slightingly, even in its mildest form, for very gradual transitions into the genuine croup happen, and, after the fatal termination of which, we may, when too late, regret having carelessly treated the first hoarseness." During the entire course of the disease, the little patient should be kept in bed, especially if the weather is cold or wet, and this should be rigidly enforced till there is no trace of hoarseness left. The atmosphere of the sick chamber should be uniform, moist and warm, and to secure uniformity a thermometer should invariably be used. The warmth and moisture may be secured by the generation of steam in the apartment, and then the affected larynx will be kept from further irritation, for the warm moisture prevents the drying of the mucous secretions of the affected parts during sleep, and the patient is spared the terrible attacks of dyspnœa which result from this inspissation. The inflamed organ should be rested. On no account should the patient speak; an ample experience convinces me that silence is often absolutely indispensible to a cure. I have seen good results from a warm compress to the throat, but only evil from the applications of pounded ice, introduced, I believe, by Dr. John Mason Good. The diet should be bland and demulcent, and if dysphagia is at all marked, nutrition by the rectum should be at once commenced.

In the prevention of acute catarrhal laryngitis the practitioner will soon discover the value of Felix von Niemeyer's advice. "It is advisable rather cautiously to habituate children to the causes of this disease, than to enervate them by a systematic over-protection, which tends to increase the liability to its attacks upon every trifling occasion." To this end, the child must not be shut up in the house merely because it has once had an attack of this disease. The little one should be in the open air every suitable day, and, when in the house, close and over-heated rooms, especially bed-rooms, should be sedulously avoided,

But, when in the open air the child should not dawddle round in the manner too often seen, but should be encouraged to indulge in active exercise. Flannel underclothing should be worn in winter and merino in summer, and the underclothing should come high up on the neck, but the neck should not be burdened with additional shawls and mufflers. F. von Niemeyer tells us that a silk ribbon worn about the neck has the reputation of a sympathetic prophylactic, and, as no harm can possibly result, it would be well to test the somewhat eccentric recommendation. Sponging the entire body with tepid salt water every morning has a very excellent effect, though if the throat and neck are washed with cold water the result is still better. Von Riemssen advises the use of the *rubbing wet sheet* of the hydropaths, and I can endorse the recommendation after a long experience of its good effects. "In such cases it is well to have the whole body rubbed every morning with a large sheet, which has been previously dipped in cold water, and carefully wrung out. As the patient gets out of bed his night linen is removed, and the sheet, which is held spread out, is thrown around him from behind, so as to cover the head, but not the face, and the whole body down to the feet. A gentle, rapid friction of the skin by rubbing with the sheet will diminish the unpleasant impression from the cold moisture. After one or two minutes of this friction the wet sheet is removed, a warm, dry one is thrown about the body in the same way, and the body is dried. The patient then puts on his clothes, and immediately takes out-door exercise, whatever the weather. If the skin be very delicate, I modify the treatment by at first giving the water, into which the sheet is dipped, a lukewarm temperature (about 86° F.), and then lowering the temperature two degrees daily, until it reaches that of spring water (50° to 56°). This treatment I have adopted for several years, with the best results for children as well as adults, and the patients never catch cold, if the rubbing be done in a warm room with the feet resting on a woolen rug. After using this treatment for eight days the patient may be allowed to wear less clothing. During the winter he may continue to use a fine woolen under-jacket, notwithstanding the frictions, but in the spring this garment must by all means be discarded, and about this time the cold frictions are to be resumed, if they are not employed both summer and winter, as is advisable in the case of children." Whenever at all practicable, children subject to catarrhal laryngitis should spend at least a portion of each summer at the seaside, for nothing diminishes the sensitiveness of the skin and respiratory mucous membrane like the exhilarating air of the sea-coast.

In almost every case Aconite is the first remedy indicated, though Bæhr remarks that "upon the whole, this remedy does not seem often indicated in simple catarrhal affections, except, perhaps, in the case of children in whom the febrile symptoms assume a different shape from what they do in full grown persons." He adds, however, that "for catarrhal croup it is undoubtedly the best remedy, which, however, ceases to be indicated if the physician be not called till the second or third day of the disease." It is emphatically *the* remedy when the disease is caused by exposure to keen, dry, cold air, and as soon as the laryngeal inflammation is lighted up, this remedy, if given in repeated doses, will often cut short the most severe attack. The skin is hot and dry, with full, quick and bounding pulse ; the voice is rough, hoarse and tremulous (with Belladonna and Bryonia the voice is nasal or raised) ; painful sensitiveness of the larynx, with aggravation on coughing or speaking ; short, dry cough, with constant irritation ; during the day the cough is short and panting, at night it is rough and hollow ; accompanying this cough there is expectoration of scanty, whitish mucus; the face and eyes are red and flushed ; great thirst is present; the sleep is broken and restless, and the entire nervous system is irritable. I have had the best results from the 4th or 5th decimal triturations of the root of Aconite, but it must prove curative in all dilutions and preparations.

Sanguinaria is indicated by dryness of the throat, with

soreness, swelling and redness ; sensation of swelling in the larynx and expectoration of thick mucus; tickling in the throat in the evening, with slight cough and headache ; dry cough, awakening him from sleep, which did not cease until he sat upright in bed, and flatus was discharged upwards and downwards; tormenting cough, with expectoration ; circumscribed redness of the cheeks; continued severe cough, without expectoration ; pain in the breast, and circumscribed redness of the cheeks. I have used Sanguinaria successfully in many cases of laryngitis, and I look upon it as being the leading remedy in this disease, Aconite not excepted. I have generally used it in the form of an acetous syrup, as directed in the chapter on pseudo-membranous croup.

Spongia is a most important remedy when croupous breathing appears in the course of the disease. Laurie says that it should, in the generality of cases, be administered six hours after the last dose of Aconite, and Bæhr thinks that it is the principal remedy for this disease when accompanied by distinct symptoms of œdema of the mucous lining of the glottis. The cough is dry, barking and hollow, coming on in paroxysms, especially at night, with shrill and wheezing breathing ; when there is expectoration it is only in the morning (with Hepar there is expectoration in the morning and during the day) ; it is improved by eating and drinking, worse when sitting erect, from motion and exertion (the cough of Hepar is worse when lying). The larynx and upper part of the trachea are painful and sensitive to the touch, the hoarseness is very marked and the patient speaks with difficulty. Bæhr remarks that Spongia is likewise appropriate, if the croupous sound of the cough still continues, and lumps of a tenacious, yellow mucus are expectorated. I have had the best results from the lower triturations from the 3d to the 6th decimal.

Hepar is very similar to Spongia, but there is rattling of mucus in the larynx from the commencement ; the cough is moist with great hoarseness, and slight suffocative spasms are present. Hepar is worse indoors and in the morning, while Spongia is better outdoors and in the morning. Laurie remarks that Hepar may be selected to follow Aconite in preference to Spongia, if the fever and burning heat of the skin continue, notwithstanding the previous administration of Aconite. I have had the best results from this remedy in the 4th or 5th decimal triturations.

Tartar emetic is indicated when there is hoarseness from the very beginning of the laryngeal inflammation, a hard and ringing cough, or paroxysmal fits of coughing, with suffocative arrest of breathing and profuse secretion of mucus. Hartmann gives this remedy "if the inspirations should evince a paralytic condition of the lungs," and in this state Tartar emetic will help unless the patient should be beyond the reach of help. I have succeeded best with the 3d or 4th decimal triturations in repeated doses.

Belladonna is, according to Dr. Duncan, of Chicago, the leading remedy in simple, uncomplicated cases of this disease, and we thank that excellent observer for pointing out a fact which, though recorded in our literature, had passed from the professional mind. Very many cases are simply congestion in the first place, and, if Belladonna is promptly given, the results are very striking. Violent stinging pains in the larynx are present, with dry spasmodic cough, coming on in paroxysms, aggravated particularly in the evening and before midnight. Croupous breathing with hoarseness is present, and the voice is low and feeble—in some severe cases this proceeds to complete aphonia. The pharynx and tonsils usually participate in the inflammatory action, and deglutition is difficult and painful. Fever is present with disposition to perspire and to sleep; the pulse is full and bounding. Laurie says that Belladonna is not to be administered in cases in which it has been previously employed, as, for instance, if the affection of the windpipe occurred immediately after an attack of pure scarlet fever. I have commonly used Belladonna from the 6th decimal to the 12th centesimal trituration, though doubtless it would be effective in almost any dilution.

Mercurius solubilis is said by Bæhr to act similarly to Belladonna in this disease, though I have never been able to see the resemblance. There is chilliness and great sensitiveness to cold, with frequent paroxysms of dry, burning heat, alternating with copious perspirations which do not afford relief; the larynx is sore, the patient is hoarse, but there is no loss of voice; the dry, distressing cough occurs principally at night, and the catarrhal inflammation extends to the eyes, nose, pharynx and even to the mouth. I have had the best results from this remedy in repeated doses of the 4th to 5th decimal triturations given dry on the tongue.

Bryonia is an excellent remedy to follow Aconite, or, in mild cases, from the commencement. The cough is spasmodic and suffocating, especially after midnight, with expectoration of yellowish mucus, hoarseness with rattling of mucus in the larynx, and tenderness of the larynx on pressure. In young children I have usually given the 12th centesimal dilution, but, on the whole, Bryonia is not so useful in children as in adults.

Phosphorus has fever with hoarseness and dry spasmodic cough, with stitches in the larynx and constriction of the throat. The voice is trembling or hissing (in Bryonia it is raised or nasal), or there may be complete extinction of the voice. I have had the best results from the 12th centesimal dilution, and it is an excellent remedy to follow Aconite.

Arsenicum album is indicated by glowing fever-heat with constant thirst; general debility with a prostrate feeling in the whole body; burning pain in the larynx, increased by deglutition, which is difficult; short, dry, hoarse cough in rapid paroxysms, with violent action of the pectoral muscles during an inspiration. Arsenicum acts well in all dilutions; I prefer the 12th centesimal.*

Lachesis is a valuable remedy when the larynx is raw and dry and very sensitive to the touch. No laryngeal spasm is present, but the patient is hoarse, with a feeling as though something had to be hawked up, and a good deal of pain and difficulty is experienced on swallowing. Lachesis is often indicated after Hepar, and I have had excellent results from the 12th centesimal dilution.

Minor remedies are Nux vomica, useful during the decline of an attack after the fever has abated, when there is still an evident sense of constriction during breathing, with a constant tickling, hawking cough with tenacious expectoration; Hyosciamus, useful when after the cure of the laryngeal inflammation, a spasmodic cough, only at night, remains to harrass the patient; Argentum nitricum, when the disease tends to assume the chronic form, with swelling of the posterior wall and lining of the larynx, with hoarseness and loss of voice, continual and vain efforts to swallow, with pain and soreness in deglutition, much hawking, considerable muco-purulent expectoration or titillation in the larynx, with dry, spasmodic cough; Pulsatilla when the patient is chilly with titillating cough, excited by a sensation of scraping and roughness in the throat, spasmodic and setting in more especially in the evening and when lying down, better on sitting up, commencing again on lying down, and sometimes increasing to suffocation; lastly, Hartmann advises Ipecacuanha or Sambucus if, after the abatement of the fever, local symptoms should still remain with anxious and hurried breathing.

APHORISMS.

1. Acute catarrhal laryngitis, also called catarrhal croup, is simply a catarrh of the larynx, which assumes the croupous form when the epiglottis and rima glottidis are involved.

2. The duration of the disease is from four days to a week, and, should it last longer than a week, there is a danger of the advent of pseudo-membranous croup.

3. Even in healthy children a simple catarrhal laryngitis, apparently devoid of danger, may be converted into true croup—one of the most serious of diseases.

4. Even in its mildest form, acute catarrhal laryngitis should never be regarded slightingly.

5. The silence of the patient is often indispensible to a cure. Rest the inflamed organ.

6. Acute catarrhal laryngitis is best prevented by gradually accustoming children to the causes of the disease, and by the judicious use of hydropathic appliances.

7. The leading homœopathic remedies are Aconite, Sanguinaria, Spongia, Hepar, Tartar emetic and Belladonna.

8. Belladonna, one of the chief remedies, has been too much neglected hitherto; Sanguinaria will remove the predisposition to the disease.

CHAPTER VI.

ACUTE ŒDEMATOUS LARYNGITIS.

This disease is not very common among children, and, as it occurs under varying conditions, many excellent writers consider it a mere symptom supervening on the morbid states, but as Prosser James remarks, " it is a condition so important as to deserve to rank separately as a disease." Its rarity is accounted for by the fact that in young children there is very little submucous areolar tissue in the larynx, consequently very little field for submucous effusion during inflammations of that organ.

Acute œdematous laryngitis may be defined to be an inflammation of the submucous areolar tissue of the larynx, resulting in infiltration of serous, sero-purulent or purulent fluid, accompanied in serious cases by stridulous breathing, orthopnœa, and dysphonia or even aphonia. The older observers considered that it was non-inflammatory in its nature—in fact a pure dropsy—but later investigations have conclusively shown that inflammatory œdema of the larynx is much more frequent than non-inflammatory infiltration. The older name—œdema of the glottis—has been gradually abandoned for the more appropriate one of œdema of the larynx, for the glottis is not specially the seat of this affection. Some writers apply the term *submucous laryngitis* to this disease, but Trousseau objects to it on the ground that it conveys the idea of an inflammatory malady, although the name he proposes—*angine laryngée œdémateuse*—is open to a similar objection. Bouilland proposed the term *laryngitis phlegmonosa*, indicating the identity of the morbid process with the phlegmonous inflammations of other mucous membranes, and von Riemssen has definitely adopted the term.

Œdematous laryngitis most frequently occurs in feeble children, and at first it may be mistaken for a 'cold'; as the disease advances, croup presents itself to the mother's mind. Children who have chronic tonsillitis are liable to it, and in such cases the disease commences in a very insidious manner. It may come on when the little patient is recovering from measles or scarlatina, and in the latter disease it has been more frequently observed in patients who have had the disease in a mild form than in those who have had a severe attack; indeed, I have come to look for it, especially in patients in whom the disease has been wholly without eruption. It may be developed during the course of albuminuria, and, indeed, in connection with any disease on which anasarca may supervene. It accompanies tuberculous disease of the larynx, and in scrofulous subjects suffering from erysipelas of the head and face, the physician should be on his guard that it does not prove a suddenly fatal complication. Sir Thomas Watson says that he has known such an inflammatory œdema to arise from a mercurial sore-throat in a broken-down constitution.

Acute œdematous laryngitis is either *primary* or *secondary*, though Dr. Paul Guttmann of Berlin opposes Sestier, Trousseau, Mackenzie, Bæhr and almost the entire profession by stating that "œdema of the larynx is invariably secondary." The disease is said to be *primary* when it attacks children previously healthy; *secondary* when it affects those already suffering from disease. Sestier, who, more than any one, is competent to speak with authority on this disease, in one hundred and ninety cases found thirty-six primary, and a hundred and fifty-four secondary. Again, the disease is *typical*, when originating in the larynx, *contiguous* when it spreads from the pharynx or other adjacent parts, and *consecutive* when it depends upon some organic disease of the larynx. Again, the morbid state may be either acute or chronic, and at times it assumes an epidemic form, and I remember an epidemic of scarlatina, marked by the prevalence of œdema of the larynx in the contiguous form—but only when the scarlatina was declining. The mechanism of the disease will readily be understood when we reflect that the submucous tissues are in a state of sub-acute inflammation, that effusion has taken place, and that the resulting swelling obstructs respiration, and, as the swelling is usually greatest in the epiglottis and upper part of the larynx, inspiration is more difficult than expiration. "Considering the manner in which the disease originates, the most correct explanation seems to be that a suppurative process in the neighborhood of the glottis causes œdema in the same manner in which a chancre causes within a few hours an excessive œdema of the prepuce." (Bæhr). Again, the disease is *supra-glottic* or *sub-glottic*, the latter being an inflammatory œdema of the parts below the vocal cords, more difficult of diagnosis than the supra-glottic variety, and, when operative measures are indicated, calling for tracheotomy rather than the scarifications so effective in the supra-glottic variety. I propose considering in this chapter the non-inflammatory form of œdema of the larynx, as well as the inflammatory, although the treatment necessarily varies with the cause of the morbid state.

None of the medical writers of antiquity describe this

disease with anything like clearness, and this is not to be wondered at when we reflect that they merely described the symptomatic appearances observed during life, and that they were almost wholly ignorant of morbid anatomy. In the year 1765, Morgagni, in his famous work, *De Sedibus et Causis Morborum*, first clearly described the post-mortem appearances, and later Boerhaave and his commentator, Van Swieten, added to the store of knowledge. In 1801, Bichat described the malady as being wholly unique—" a particular kind of serous swelling that does not occur in any other situation"—from which it is evident that he did not clearly understand the mechanism of the disease. In 1808 Bayle presented to the Medical Society of Paris his *Memoire sur l'œdème de la glotte ou angine laryngée œdemateuse*, which constitutes the starting point of a scientific knowledge of this disease. I have not had access to the original, and authors differ as to the true nature of his views, for, while von Riemssen states that " Bayle's œdema glottidis is a serous infiltration of the submucous connective tissue, non-inflammatory in its origin," his countryman, Trousseau, says, " I repeat, that you will almost constantly see œdema of the larynx depending on inflammation, a fact which Bayle established and was the first to describe." Finally, in 1852, Sestier gave us a standard work, including references to 245 cases, not including cases of scald-throat. Still later, Gibb, Mackenzie and von Riemssen have systematized our knowledge, and at the present time the disease is almost as well understood as any other laryngeal affection.

The disease is somewhat rare in childhood, though there is reason to believe that many fatal cases are attributed to other diseases. In 215 cases Sestier found only five in children under five years of age, one of them being a newborn infant, and twelve cases between five and fifteen years. Again, in 245 cases, Sestier noted only two primary cases in children, and in two other cases, between the ages of four and six years, the disease arose by propagation from inflamed neighboring parts.

Acute œdematous laryngitis may supervene on a slight attack of local inflammation, as catarrhal pharyngitis, or, more frequently, erysipelas of the pharynx, though, as a general rule, it follows deeper seated affections of the larynx. Indeed, Sestier asserts that "simple inflammation" is the cause of œdema of the larynx in only six per cent. of all his cases, while in twenty per cent. of his cases propagation took place from the pharynx; and the pharyngitis was in many cases moderate and even slight.

Œdematous laryngitis almost constantly commences with a chill, even when it appears as an intercurrent disease. The chill alternates with flushes of heat, and soon the skin is hot, the pulse full and bounding, and the face red and flushed. Deglutition is difficult, partly from the pharyngitis so frequently present, and partly from the swollen epiglottis permitting food to enter the larynx. The external parts are swollen in both the primary and secondary varieties of the disease, and the swelling is simply the serous effusion following sub-acute inflammation. The patient complains of sore throat, and one of the misleading features of the malady is that, on examination, the tonsils and pharynx appear to be the seat of the disease. Often the patient, if old enough to describe his feelings, complains of a pricking, burning pain in some particular part of the larynx, and this pain is increased by deglutition and accompanied by a slight, irritative cough without expectoration. But soon the voice gets rough and hoarse, and this rapidly-increasing hoarseness soon passes into almost complete aphonia. At the same time complaint is made of pain as if a piece of wood or other foreign body were wedged in the larynx, and this gives rise to repeated efforts, by swallowing or by coughing, to clear the throat of the offending substance. But the harsh and painful cough only results in the difficult expectoration of a little viscid mucus, which brings no relief. The most prominent symptom, however, is an impediment to respiration, sometimes increasing gradually, at other times with such frightful rapidity that the fatal result takes place almost immediately. At first the respiration is whistling and wheezing, at an advanced stage it is rasping and sawing. At the commencement of the morbid state the oppression is greatest during inspiration, which requires considerable effort and is accompanied by a kind of snoring noise, especially during sleep. Expiration, of course, is performed more readily than inspiration, for the swollen membrane closes like a valve against the entrance of air, but readily permits it to pass out. As the disease advances, however, expiration, hitherto easy, noiseless and hardly perceptible, becomes difficult and the dyspnœa rapidly increases. There is, however, even in cases in which permanent obstruction is present, the same tendency to remissions and exacerbations that characterizes almost all affections of the larynx, though the paroxysms may last as long as ten to fifteen minutes, during the whole of which time suffocation appears to be imminent. During the paroxysm the patient stands up from the bed and instinctively makes for the window with mouth wide open, agonizing for breath. The face is livid and cool, the nostrils are distended, the eyes seem to start from the head, the whole body is trembling and convulsed, and the skin is bathed in perspiration, the cough becomes less frequent, and as the disease advances it disappears altogether, as the patient cannot inflate the lungs. As the carbonic acid poisoning advances the face becomes of a dusky, bluish hue, delirium appears, during which the sufferer tears at his neck in a kind of frenzy, and unless prompt relief is given he dies strangled.

Should a favorable change take place, all the symptoms abate; the difficulty of breathing diminishes, though it still remains somewhat embarrassed, especially during inspiration; the cough becomes easier and more sonorous; the voice once more becomes audible. It is some time before the little patient recovers from the effects of such a storm, and he must be carefully guarded against relapses.

From first to last the disease, in its primary form, may last from three to five days; as an intercurrent disease, it may prove fatal in a few hours. Bæhr says that the affection may last from twelve hours to upwards of a week, and von Riemssen holds that highly-acute diffuse infiltration runs its course under the most stormy manifestations, and may cause death within a few hours or even minutes, through closure of the laryngeal entrance, if the right help is not afforded at the right time. Prof. George B. Wood thinks it probable that life is sometimes suddenly terminated by the supervention of spasm of the glottis, and the writer has seen cases in which this has actually occurred.

Scald-throat, as it is popularly called, is a very common and very fatal form of œdematous laryngitis among the children of the laboring classes in England, and indeed in every country in which tea is an ordinary beverage. The accident usually happens to young children in the mistaken attempt to drink boiling water from the spout of the tea-kettle. The boiling water rarely reaches the œsophagus, for it is expelled by the spasmodic action of the muscles of the pharynx, but it has had time to come in contact with the inside of the mouth, the epiglottis and aryteno-epiglottidean ligaments. Probably the screaming, caused by the acute pain, causes a sudden inspiratory effort which draws the boiling water, and still more readily the heated steam, towards the larynx. Even when the larynx itself is not scalded, it soon becomes involved by extension of inflammation from the pharynx. Sometimes a deceitful calm of an hour or two follows the scald, when suddenly the laryngeal symptoms are developed, and the state is at once alarming in the extreme. More commonly, hoarseness and dysphagia appear at once, accompanied by inflammatory fever, and this is followed in a few hours by œdema of the larynx, with difficult inspiration, hoarse, croupous breathing, and even spasm of the glottis. The morbid state marches on with frightful rapidity, the face becomes bluish, the hands and feet cold, the breathing more and more oppressed, the voice becomes extinct, and death takes place by suffocation.

The post-mortem appearances in cases of scald-throat are those of intense inflammation of mucous membrane of all the affected parts, and especially the mucous membrane of the epiglottis and of the aryteno-epiglottidean ligaments is thickened from effusion into the sub-mucous areolar tissue. Sometimes the opening of the larynx is quite closed, but the œdema never extends below the vocal cords. When death is very speedy, and speedy deaths are very common here, the mucous membrane below the rima glottidis may be quite normal, but when the patient has survived for some days the trachea and bronchial tubes are inflamed, and the lungs congested, or even hepatized.

Not even Wunderlich, with his tireless industry, has given us any observations as to the temperature of the body in this dreaded disease, and personally, when I had a case, I thought more about the lancet than the thermometer.

The inflammation present in œdematous laryngitis is of a low grade, and the effusion is produced in the sub-mucous areolar tissue as the result of inflammatory action in that membrane, or as the result of inflammation in adjoining parts, as Bæhr has so well pointed out. As a general rule, the affected parts have the usual red hue of inflammation, but about the entrance to the larynx they are often transparent, fluctuating and of a pale-yellow color, especially the parts well supplied with areolar tissue, as the duplicature of the aryteno-epiglottidean ligament. These bulge out in two loose and pendulous rolls extending backwards to the pharynx, while the bloated epiglottis projects high above the root of the tongue. Sometimes but one of these ligaments is affected, and a single transparent swelling closes, more or less, the entrance of the glottis. Sometimes, but rarely, the sub-mucous tissue of the vocal cords is affected, and the sub-glottic form of the disease, so well described by Sir George Duncan Gibb, is almost as rare, simply because, as he points out, "the sub-mucous tissue at the upper part of the larynx is loose, and quickly admits of infiltration and swelling, or œdema, during inflammation ; but below as well as in the trachea, it is less in quantity, and of a more dense quality, therefore, inflammation is not succeeded so rapidly by sub-mucous effusion as it is by exudation of lymph upon its surface." At times the œdema extends down the trachea, but Sestier detected it only seven times in 132 cases of œdema of the upper air passages. Very commonly, the neighboring muscles are saturated with the serous or sero-purulent fluid. Generally the effused fluid is sero-purulent, for pure serum is found only in the *foudroyante* cases; according to Sestier blood is often mingled with serum in that very class. In the more chronic cases the fluid is quite purulent.

The diagnosis of œdematous laryngitis is surrounded by difficulties, yet much depends upon the disease being recognized at an early stage. Some physicians include all laryngeal diseases of children under the generic name of "croup," and such a wholesale ignoring of pathology must result in a largely-increased mortality in this class of diseases. M. Thuillier's test, insisted on by all practitioners who have studied the disease, is almost decisive as to the existence of the supra-glottic variety, though, for anatomical reasons, it affords us little or no help when the disease is sub-glottic. By simply depressing the tongue, the epiglottis rises as a pale or reddish, pear-shaped swelling behind the root of the tongue. Then when the index-finger is rapidly but gently passed into the larynx, the œdematous swelling can be distinctly felt. With one hand the physician should press up the os hyoides so as to bring the glottis more within reach, while the forefinger of the other hand is engaged in exploration, but, as already remarked, we can only detect the œdematous swelling of the epiglottis and aryteno-epiglottidean ligaments. Dr. George B. Wood thinks that this mode of examination must be difficult, and that it might possibly aggravate the inflammation, but in practice one almost always succeeds, though with varying facility. Trousseau observes that exploration by the finger must be practiced in a very careful manner, and he adds that while he was examining the throat of a woman in the most guarded possible way, he induced a suffocative seizure, which very nearly proved fatal. In adults the laryngeal mirror would aid much in the diagnosis, but it is difficult to use it in children.

Acute œdematous laryngitis is very likely to be confounded with croup, and in very many points there is a very strong resemblance. It resembles croup in the difficulty of breathing, the suffocative fits and the cough, the hoarse voice, and the noisy, stridulous inspiration, and even in the intermissions between the paroxysms, which, indeed, are common to all laryngeal diseases. But œdema of the larynx chiefly occurs as an inter-current disease in children suffering from some malady of the adjacent parts, while croup almost always attacks children in good health. Again, the cough of œdematous laryngitis has not the croupous, brassy sound of the cough of croup. In œdematous laryngitis the difficulty of breathing is greatest in inspiration, while expiration is comparatively free, but in croup of any kind inspiration is as difficult as expiration. Lastly, œdematous laryngitis has no exudation in the pharynx and no expectoration of membranous shreds as in true croup.

The sub-glottic variety is distinguished from the supraglottic by the absence of the shrill whistling inspiration so marked in œdema of the upper part of the larynx, and on examining with the finger the epiglottis and aryteno-epiglottidean folds are normal or nearly so. Sir G. D. Gibb points out that the effusion in the sub-glottic variety is "invariably fibrinous," never serous as in the supra-glottic form, and he says that it may be taken as a curious and undisputed fact that the sub-glottis, from its anatomical peculiarities, secretes fibrin which may be poured out on the surface of the membrane or beneath it, according to the special exciting circumstances inducing it. It is undoubtedly true that the parts below the glottis are less abundantly supplied with sub-mucous areolar tissue than the supra-glottic region, and also that, as a general rule, the tendency of that part of the larynx, when inflamed, is to throw out fibrin, but Burow, Rauchfuss and Lefferts of New York have all reported unquestionable cases of sub-glottic œdema in children.

Mackenzie says that all the examples of sub-glottic œdema he has met have been of a chronic character, but, curiously enough, all the present writer's cases were acute, similating membranous croup very closely indeed.

Œdematous laryngitis is always a most serious disease, even when recognized at its inception. Indeed, the prognosis is favorable only when the grade of the disease is not at all marked and when the inducing cause has ceased to progress, or when the œdema affects only one aryteno-epiglottidean fold or but one side of the epiglottis. Bæhr says that "the most common termination is death by suffocation, and the prognosis is consequently that of inevitable death;" von Riemssen teaches that "the higher grades of laryngeal stenosis, due to sub-mucous infiltration, usually terminate in *death* if timely interference does not prevent," while Prosser James states that "if not relieved it will be fatal in a few hours, and cases are recorded in which no warning preceded death, which, therefore, may be termed sudden."

Bayle, writing when the mechanism of the morbid process was but little known, reports seventeen cases with but a single recovery. Sestier compiled statistics of almost all the authentic cases on record, and the mortality was 158 in 213 cases, though the trachea was opened in thirty of the fatal cases. In the 55 recoveries, tracheotomy was performed twenty times. The primary form of the disease is less dangerous than the secondary, and typical œdema—the form originating in the larynx itself—is almost invariably fatal. "When œdema of the larynx is a primary affection, or is connected with acute inflammation of the pharynx or larynx, its progress is more rapid, and the chances of a favorable termination are also greater, which arises from the affection being transient in its nature like the pathological state on which it depends" (Trousseau). If the inflammatory action should originate in the pharynx, the prognosis is comparatively favorable, but if it commences in the areolar tissue of the neck it is almost invariably fatal. Still, one of the worst cases the present writer ever saw, commenced in the areolar tissue of the neck, and was cured by Sanguinaria, as detailed in the remarks on therapeutics. If the disease has its starting point in syphilis, as is not seldom the case with children, the morbid state is generally curable, but if it should occur during the course of typhoid fever, the case will likely prove fatal. The supra-glottic form is more dangerous than the sub-glottic, simply because the parts above the glottis are richest in areolar tissue. It is, of course, less dangerous in a child of good constitution than in one of scrofulous diathesis or in feeble health. Habitual disease of the larynx would materially darken the prognosis, and when the disease supervenes upon chronic laryngitis, it is almost incurable. In the advanced state, when asphyxia has already commenced, there can be but little hope, for, even if the dyspnœa is relieved, the nervous system may be unable to rally from the prostrating influence of the poisoned blood.

Speaking of laryngitis, Dr. Richard Hughes observes, "should œdema glottidis supervene, repeated doses of Apis would give the best chance of averting tracheotomy," and in his latest work he further says, "It (erysipelatous sore throat) is often the beginning of œdema glottidis, in which Apis is the great remedy. It has proved curative in more than one instance of this affection, where the cause was drinking water from a kettle. Such cases are commonly fatal." Again, in his excellent *Manual of Therapeutics:* "I think that the best advice I can give you as to the treatment of this dangerous condition (œdema glottidis), under whatever circumstances it may occur, is to trust to *Apis*. Since this remedy has cured it even in its most fatal form—viz., that which occurs in children after drinking from the spout of a tea-kettle—it will probably be competent to deal with all other forms of the malady." Dr. Holcombe observes that Apis is especially indicated when the attack has suddenly sprung up in the course of an acute disease, in otherwise healthy persons, and that it is still more so when it occurs n erysipelas, burns, or the eruptive fevers to which the bee-virus has more or less affinity. Bæhr regards it as one of the three remedies which act similarly to the general disease—the others being Lachesis and Rhus toxicodendron. I look upon Apis as being undoubtedly the first remedy to be thought of, though, of course, it does not cover all cases. Still, I have succeeded with it when success seemed to be unattainable. Holcombe recommends the 3d dilution to be used; I always use the 5th decimal trituration of the sting of the bee and the attached sac of the virus, for, as Constantine Hering remarks, "there is but one right kind," and that is it.

An English physician, Dr. Ainley, of Halifax, communicates the following excellent case to the *Homœopathic World*, vol. XIV:

"In November, 1878, I was summoned at 11 P. M. to see a little boy, aged four years, who had been taken ill. The history of the case was that he had been all right up to teatime, and, indeed, on being put to bed at 8.30 appeared the same, but on being looked at by the parents before they retired to rest, as was their custom, they found him breathing very heavily, and were alarmed and sent for me. When I arrived, in a moment I diagnosed "croup"—that is to say without asking any questions—and seeing no time was to be lost, as the boy's face was already blue and swollen from impeded respiration and deficient aëration, I began to prescribe my usual remedies, and which I am thankful to say usually succeed, viz.: *Aconite* and *Spongia*, administered every ten minutes in alternation. But as I anxiously watched the case, feeling sure a short time would decide for or against, I entered into conversation with the parents, and began to make fuller inquiries into the previous history of the child, and the following little incident was told me, which of course turned the whole case: On the same day, at teatime, when the mother had just filled up the teapot with hot water, and left it on the edge of the table, the little fellow drank out of the teapot-spout, and although it was very hot, he seemed to make no complaint of any pain in his throat, and played for some time, and even went to bed without complaining. Here we had an entirely new condition of things, which could have had no true interpretation apart from the incident just related ; symptomatically it was a case of " Cynanche Trachealis ; " pathologically it was "Cynanche Trachealis ;" and I suppose if one had searched through all the homœopathic literature extant only one medicine could have been found to have met the case, and that was *Apis*. *Apis* was promptly given, and in from four to six hours all danger might be said to be over."

Dr. Bruckner, of Basle, publishes an interesting case in the A. N. Z., 1873, of which the following epitome is given by Raue in the *Annual Record* for 1874 : " A young man, who had scarlet fever as a child, suffered from that time from an œdematous swelling of some part of his body, regularly returning every eight days. For the last three years it threw itself sometimes on the glottis, causing fits of suffocation, but always terminating in twelve hours. Before the paroxysm attacks of bilious vomiting. Relieved, but not cured, by Apis 200."

In 1869 I wrote : " I have never had an opportunity of testing the virtues of Sanguinaria in this disease, but would expect considerable from it ;" and in the month of April, 1874, I had the long-looked for opportunity. On Friday, April 17, 1874, I was called to Mrs. C., aged 59, who had been complaining for some few days. I found an inflammation of the cervical glands of the right side, involving the parotid gland to a considerable extent, and accompanied by extensive inflammation of the subjacent cellular tissue. The parts were hot, tender, swollen and red—in fact, the well-known *calor, dolor, tumor, rubor*—and there was reddening of the fauces, with slight pain on deglutition. I prescribed Belladonna, 6th decimal trituration, and advised rest, quiet and silence. On the following day the situation was but little changed, and Mercurius iodatus ruber, 3d decimal trituration, was prescribed.

At 6 o'clock of Sunday morning, April 19th, I received an

urgent call to the patient, who, I was told, had hardly been able to breathe all night. I found her sitting up in bed, with a characteristic rasping and sawing sound issuing from the larynx, a sound somewhat difficult of description, but which once recognized can never be forgotten. The tonsils and pharynx were swollen, but auscultation showed that the sawing and rasping sound issued from the larynx. The cough was dry and harsh, relieved by sitting up in bed, aggravated by eating and lying down, and it was accompanied by difficult expectoration of tough and glairy mucus. The voice was low and suppressed, and it was with difficulty that I could make out the hurried, whispered sentences.

The pulse was feeble and fluttering, and the lips were pale; but on both cheeks there was a circumscribed redness. The pathognomic symptom which made the pathological state quite clear to me was the fact that expiration was performed more readily than inspiration. M. Thuillier's test was decisive as to the diagnosis, for " when the forefinger was passed into the larynx, there is a perception of a cushion formed by the tumefaction of the sides of the glottis—a soft, pulpy body, quite distinct from the ordinary hard feel of the parts."

The diagnosis was acute œdematous laryngitis of the supra-glottic variety—all the more dangerous because it was an intercurrent disease—and the peculiar respiration arose from the fact that the œdematous membrane which filled the glottis closed like a valve against the entrance of air, but readily permitted it to pass out. I prescribed Sanguinaria 1st decimal trituration, a dose every half hour.

At 1 P. M. I found that improvement had commenced almost as soon as the medicine was given. The sawing and rasping sound was now much diminished, the respiration was comparatively easy, inspiration and expiration were performed with equal facility, the cough was less frequent and less severe, the voice was quite audible, and the patient had slept much of the time since morning. The tonsils and pharynx were still red and swollen, but the glottis was clear of the tense and rounded swellings present in the morning. The Sanguinaria was continued in the same dose.

At 7 P. M. I again saw the patient and found that the very serious pathological state had almost wholly disappeared. The Sanguinaria was continued all night, and in the morning, as the acute œdematus laryngitis was no longer present, treatment was directed against the inflammation of the cervical glands and cellular tissue."

"Should Apis fail you, however, you may (before thinking of surgical measures) consider the claims of Sanguinaria." (Hughes.)

In 1869 I suggested Aconite as a leading remedy, and, although no other writer of our school, save Charles Julius Hempel, has endorsed the recommendation, I repeat the suggestion with all the more confidence that I have found its action prompt and decided in several well-marked cases. But it must be given in repeated doses of the mother tincture, or 1st decimal dilution.

Dr. Jacob Reed, Junior, of Philadelphia, reports the following case: "March 16, 1867, evening. Called to see Miss B., at 20, who had for some days 'had a bad sore throat,' and was reported as choking to death. When seen, the patient was evidently suffering from an acute œdematous inflammation of the larynx, there being high fever, pain in the region of the larynx, difficulty of swallowing and breathing, voice almost inaudible, every effort at speaking causing great pain, inspirations prolonged and stridulous, being effected only by violent effort; there was but little cough; frequent spasmodic exacerbations of these symptoms rendered suffocation imminent. Ordered inhalations of steam medicated with Opium, cold pack to the region of the larynx, Aconite and Kali bichromicum; of the Aconite three drops of the tincture of the root were given in a half glass of water, of which she took a teaspoonful every twenty minutes. This appeared to afford relief, which, however, proved but temporary, as upon paying my morning visit, I found the patient much worse in every respect, the leaden hue of the skin, with the intense anxiety of the countenance, showing that she had to fear the result of deficient aeration of the blood. This condition of affairs rendering tracheotomy necessary, I returned to the office for the necessary instruments and assistance, but in the meantime ordered two drops of the tincture of the Aconite root to be given every ten minutes. Upon returning after the lapse of an hour, the patient was so far relieved as to render surgical interference unnecessary, and from this the convalescence was steady, although slow and imperfect. There remains, after many months, a cough, with hoarseness, owing to constitutional tuberculosis."

According to Bæhr, "we are acquainted with only one remedy which has œdema of the glottis among its physiological effects; that remedy is Iodium." Holcombe, too, advises it and I look upon it as being one of our chief remedies. In addition to the administration of the remedy in the ordinary way, I apply the 1st or 2d decimal dilution directly to the œdematous parts.

Dr. Holcombe says that Arsenicum album is indicated when the disease is a genuine anasarca, coming on slowly in the chronic diseases of broken down constitutions, especially if there is concomitant cardiac or aortic lesion, Bright's disease of the kidneys, anæmia or dropsy. Though this remedy is also recommended by Raue, I have never seen the results that one might expect, even when it seemed well indicated. Holcombe recommends it from the 3d to the 30th dilution.

Raue and Holcombe both suggest Lachesis, and Bæhr points out that it specially has the peculiar serous infiltration of internal as well as external parts of the body, which sets in without any symptons that might properly be called inflammatory, and which reaches its full development in a few hours. I have had no experience with Lachesis.

Bæhr says "Spongia is the principal remedy for the so-called catarrhal croup with distinct symptoms of œdema of the mucous lining of the glottis," and the same distinguished writer remarks that another remedy which offers some resemblance is Phosphorus; in this case, however, the resemblance is limited to a single symptom. Holcombe thinks that Chelidonium has "some pathogenetic resemblance to many symptoms of this formidable disease," and Rhus toxicodendron is suggested by Bæhr as acting similarly to the general disease, but, so far, these recommendations have not been acted on. Raue thinks that China and Stramonium are, perhaps, the most important remedies, and that the first-named remedy would be of special value when the œdema is a so-called pure dropsy, and he further suggests Arum triphyllum, of which I have had no experience.

But let us suppose that in a case of undoubted œdematous laryngitis, the patient gets rapidly worse in spite of the best selected remedies, or that the disease was far advanced before medical assistance was called. What will the physician do in either of these contingencies? Will he permit his patient to die, or will he make an effort to remove the mechanical obstacle which impedes respiration? It appears to me that no conscientious physician of our school could possibly ignore surgical procedures, even if they had not been advised by Hughes and Hartlaub. Bæhr, too, says that "since in this disease we cannot fall back upon experience for a positive knowledge of the curative action of drugs, it would be criminally indiscreet to depend exclusively upon internal treatment." Holcombe, of New Orleans, teaches that scarification of the infiltrated tissues is of immense benefit when it can be thoroughly done, and he adds that "tracheotomy is the last, but frequently imperative resort." The surgical procedures are two in number—scarification and tracheotomy—the former of use in the supra-glottic form, the latter in the sub-glottic variety. To M. Lisfranc is due the credit of introducing scarification; Dr. G. Buck, of New York, revived the operation, and it has been still further improved by Sir George Duncan Gibb. It is recommended by all the best authors, Sestier, Valleix, von Riemssen, though Mackenzie says doubtfully that "scarification is often successful when the disease is circumscribed." In some few cases the laryngeal mirror may be employed, but in most cases the practitioner must be guided by the sensation of the finger. Mackenzie's laryngeal lancet is decidedly the best and safest instrument, though Buck's laryngeal knife is little inferior, and a common bistoury, wrapped with sticking-plaster almost to the point, is a good instrument in good hands. The older surgeons advised numerous small incisions, but von Reimssen recommends the operator to make several long incisions, whereupon the swelling generally collapses at once. Trousseau confesses that he has not had the courage to practice this operation and he considers that Buck has exaggerated both its utility and facility. Legroux recommended that the œdematous swelling be lacerated by the nail of the index-finger cut to a sharp point, but it is doubtful if the advice has ever been acted on. After the operation, warm gargles or the inhalation of the steam of hot water will encourage the evacuation of serum. Sir G. D. Gibb recommends the introduction of a suitably curved bougie, half an inch in diameter, into the larynx, for the purpose of squeezing out the serum through the punctures made by scarification, but, though this would be easy in adults, it would be difficult in children.

We have a great consensus of the authorities as to the value of tracheotomy in this disease, and here the homœopathic writers, Bæhr and Holcombe, are one with Trousseau, von Riemssen and Mackenzie, all the great lights of the other school of thought. Bæhr says that in this disease, much sooner than in croup, success may be expected from tracheotomy, for the reason that the trachea is not usually involved, while von Riemssen urges that we must bear in mind, as a general rule, that in severe cases the danger to the life of the patient, if the physician maintains an expectant attitude, as uncommonly great, and that even postponing tracheotomy for a few hours may be destructive of the patient, if the physician leaves him in the meantime, and he further points out that there is no estimating the rapidity with which stenosis of the glottis may advance. "We should make it a rule, under no circumstances to leave a patient with laryngeal œdema, and, if the instruments are not at hand in time, to perform tracheotomy with a penknife rather than let the patient suffocate. This was done by a physician with whom I am acquainted, who on making a journey across country on the island of Rügen, and, being called into a farm-house to see a patient with œdema of the glottis, found himself without even a pocket-case. The instance which Stannus J. Hughes narrates is also a very pretty illustration of this. A student of medicine saved a man, who was at the point of suffocation from œdema of the glottis, by cutting through the crico-thyroid membrane with his penknife, and introducing the tube of his penholder as a canula (von Riemssen).

But tracheotomy should not be delayed till the patient is all but moribund, and it should be persisted in, as Durham points out, even though the difficulties attending the operation are great and the chances of a successful result appear small. If the suffocative paroxysms are severe, if they follow each other rapidly, if the difficulty of breathing in the intervals of the seizures is considerable, *then* the operation should be performed at once, especially if the slightest signs of poisoning by carbonic acid manifest themselves. It is in the child a comparatively simple operation, and, while it may be the means of saving life, it never can be the cause of death. Professor Wood remarks that well-authenticated cases are on record, in which patients have been restored after respiration had ceased, and the pulse could be no longer felt at the wrist. One would think that chloroform would be exceedingly unsafe, but experience proves that it is not so, and it would be almost impossible to operate on young children without it. The patient should be placed on a lounge with a cushion behind the neck and shoulders, so that the head is thrown back and the trachea is well forward. With lamp-black or a piece of charcoal the operator should trace on the skin the outline of the incision he proposes to make. The skin is then raised and cut through, next the muscles are carefully incised and retracted with a hook on each side. The wound should be sponged before each cut with the bistoury, and all hæmorrhage should be arrested before the trachea is opened. When the white rings of the trachea are exposed, a small puncture should be made in them, which should be enlarged with a probe-pointed bistoury till the orifice is say three-quarters of an inch in length, and it is important to note that the trachea must be cut exactly in the middle line. A double canula should then be placed in the wound by means of a dilator, and the canula should be secured by means of tapes fastened behind the neck. Bretonneau lays down the practical rule that the canula should always be at least of the diameter of the glottis of the subject. After the operation, the patient should be enveloped in a warm and moist atmosphere, but, at the same time, ventilation must be maintained. Then, well-selected remedies should be administered with the view of acting on the œdematous parts. For fuller particulars on tracheotomy, I would refer the reader to the very able article on that subject by Arthur E. Durham, Assistant Surgeon to Guy's Hospital, in Holmes' *System of Surgery*, or the article in the Internation Encyclopædia of Surgery.

Mackenzie says that ice should be "uninterruptedly swallowed," and Holcombe has found it beneficial; von Niemeyer relates that under this treatment he once witnessed the recovery of one of his colleagues, in whom suffocation seemed so imminent that the medical attendants hardly dared to defer tracheotomy.

APHORISMS.

1. Acute œdematous laryngitis is not common in children, simply because in children the larynx is scantily supplied with sub-mucous areolar tissue.

2. The older writers held that this disease was non-inflammatory, but later observers have conclusively shown that

inflammatory œdema of the larynx is much more frequent than non-inflammatory infiltration.

3. Acute œdematous laryngitis is very like croup, but in the first-named disease dyspnœa is greatest on inspiration, while expiration is comparatively free, but in all the croups, inspiration is as difficult as expiration.

4. Formerly it was believed that the effusion of sub-glottic œdematous laryngitis was invariably fibrinous, but it is now quite certain that it is often serous.

5. Acute œdematous laryngitis is a very fatal disease, Sestier reporting 158 deaths in 213 cases, though tracheotomy was performed in 30 of the fatal cases.

6. Acute œdematous laryngitis originating in the larynx itself is almost invariably fatal.

7. The leading homœopathic remedies are Apis mellifica, Sanguinaria, Aconite, Iodium, Arsenicum album, Lachesis and Spongia. Minor remedies are Phosphorus, China and Rhus toxicodendron.

8. As a last resort, scarification is of great value in the supra-glottic variety, and tracheotomy in both supra-glottic and sub-glottic forms of the disease.

9. Durham urges that tracheotomy should be persisted in, even though the difficulties attending the operation are great, and the chances of a successful issue appear small.

10. Mackenzie, Holcombe and von Niemeyer all strongly advise the uninterrupted swallowing of small pills of ice.

CHAPTER VII.

SPASMODIC CROUP.

Croup is one of the most dreaded of infantile diseases, and it is also one of the least understood. There are two varieties of croup proper, the *spasmodic* and the *pseudo-membranous*, the first a severe but comparatively inocuous disease, the second apparently less severe but in reality one of the most terrible of maladies. But it must be distinctly understood that while distinct types of these maladies exist, that frequently they shade off and run into each other in such a manner that even the most experienced physicians are, at times, perplexed. A case will commence as spasmodic croup and apparently be progressing finely; when the dreaded pseudo-membranous complication makes its appearance, and soon the patient is hopeless. Or a child will have repeated attacks of spasmodic croup, recovering from each attack after a good deal of suffering; long success lulls the watchfulness of the mother, and at length an attack assumes the pseudo-membranous form, and being met by unsuitable treatment, it soon proves fatal. The name of croup conveys very different ideas to different minds, and a case which one physician dignifies with that title appears to another altogether beneath his notice. Many years ago I was visiting a physician, and as we sat gossiping in his office, he suddenly remarked that he must go and see a case of croup. Having been accustomed to see severe forms of the disease, I started up, seized my hat, and made ready for a rapid march. My friend remarked that there was no need of haste, and so, after a very leisurely walk, we came to the house. Ushered into the parlor, we found a couple of ladies sewing and chatting, and two or three children playing on the floor, but, to me, no signs of a croup patient. My friend, however, called a little child from its play and auscultated its larynx carefully, requesting me to do the same. I did so, and after a careful examination I found that the child had a very slight *cooing* in the larynx, but no cough, no hoarseness, no fever, *no croup*.

Like some other infantile diseases, spasmodic croup has been burdened with a multiplicity of names. It has been called false croup and pseudo-croup in contradistinction to true croup, commonly called pseudo-membranous croup. Guersant calls it stridulous laryngitis ; Bretonneau names it stridulous angina ; while Millar and Simpson speak of it as the acute asthma of infants. Cullen's name, "cynanche trachealis," is wholly wrong, as it directly leads to erroneous ideas as to the site of the disease, and Morell Mackenzie, the latest writer on laryngeal diseases, gives spasmodic croup as one of the synonyms of spasm of the glottis. The French writers, Rilliet and Barthez, and the American writers, Meigs and Pepper, concur in calling it spasmodic laryngitis, while Professor Wood calls it catarrhal croup, Wichmann, Michaelis and Double style it spasmodic croup, and I prefer that name, as it appears to me that the laryngeal spasm is the essential feature of the disease, while the catarrhal symptoms are less characteristic ; but it is well to bear in mind the fact that severe catarrhal or even frankly inflammatory symptoms may arise in the course of the disease, calling for modifications in treatment.

Spasmodic croup, then, may be defined to be a laryngeal disease almost peculiar to infancy, consisting of a violent spasm of the interior muscles of the larynx, combined with a catarrhal inflammation of the adjacent mucous membrane, but without pseudo membranous exudation ; this combination of laryngeal spasm with catarrhal inflammation causing important changes in the respiration and in the voice. There are thus several elements in the disease, for the nervous system is involved as well as the vascular, so that spasmodic croup is allied to the neuroses as well as to the inflammations. At times the catarrhal inflammation is quite trifling, while the spasmodic action is distinctly marked, or the inflammatory action may be very severe, with very little laryngeal spasm, in which case the disease would approximate to catarrhal laryngitis. "The spasm of the laryngeal sphincter seems to be the result of a disordered action of the excito-motor innervation of the part, the irritant, which is productive of the morbid innervation, being, in all probability, the inflammation of the laryngeal mucous membrane, which, as has been already stated, constitutes one element of the malady. The nervous element predominates in the early part of the attack, but towards the conclusion the spasmodic symptoms disappear entirely, and we have left only those which depend on the local tissue changes." (Meigs and Pepper.) Dr. Copland writes: "The experiments of Schwilgue, Jurine, Albers, Schmidt and Chaussier, as well as pathological observation, prove that the form of disease called false croup by the above authors proceeds from a similar state of morbid action as that denominated the pure disease (pseudo membranous croup), and is merely a modification resulting from less intensity of the inflammation, peculiarity of the temperament and habit of body, the causes occasioning it, and the greater predominance of the spasmodic or nervous states." This is decidedly erroneous, for spasmodic croup differs radically from pseudo-membranous croup, and I hold with Meigs and Pepper that they are distinct affections, which may, in the great majority of cases, be distinguished from each other at a very early stage by a casual observer. I concede, of course, that pseudo-membranous croup may be developed in the course of spasmodic croup, and the practical physician must never forget the pregnant words of Rindfleisch, "*the development of a false membrane is connected in the closest manner with the catarrhal state, and constitutes an anatomical acme of the morbid process.*" At one end of the scale you have a mild form of the disease, differing from catarrhal laryngitis only in the presence of a slight degree of laryngeal spasm; at the other end you have a severe type resembling pseudo-membranous croup so closely as to try the acumen of the keenest observer. Yet the distinction between the worst case of spasmodic croup and even the very mildest case of pseudo-membranous croup is of vast moment to the patient, since the prognosis is so widely different in these two diseases.

Spasmodic croup appears to be more frequent on this continent than the pseudo membranous variety, while the contrary seems to be the case in Europe. For one case of the pseudo-membranous we meet with at least ten of the spasmodic; hence, while with us in all varieties of croup massed together the mortality is comparatively small, European writers state that almost one-half of those attacked die. Spasmodic croup, again, is, generally speaking, a disease of infancy and early childhood, while pseudo-membranous croup often attacks those of maturer years.

In common with the more dangerous form of croup, spasmodic croup affects male children more frequently than female, even when the same care is taken of the patients, a circumstance of which no adequate explanation has yet been given. Out of a hundred cases, sixty will occur in boys and forty in girls, and this observation has been repeatedly confirmed. It is most frequent in fall and winter, and also in spring when winter is breaking up, and it is rarely seen in summer. As a general rule spasmodic croup is prone to appear on the banks of lakes and in the vicinity of large bodies of water.

Spasmodic croup is essentially a disease of infancy and childhood. Guersant says that it occurs most frequently between the ages of two and seven; J. Lewis Smith thinks that it ordinarily occurs between the ages of two and five; Condie has met with it in children of nine or ten months, but less frequently than in those between two and eight years. Meigs and Pepper state that "it occurs most frequently during the period of the first dentition, being more common in the second year of life, which is the time of the greatest activity of the first dentition, than at any other age, though it is often met with in the third and fourth years." Rilliet and Barthez are of opinion that it is most common between the ages of three and five, thus omitting the very year, the second, in which it is most frequently seen. A few of my cases, not more than eight or ten per cent. of the whole number, occurred during the first year of life; at least a third of the whole number were in the second year; and a somewhat smaller proportion, say one-fifth, were in the third year, after which they decreased, till in the seventh and eighth years very few cases were seen.

Spasmodic croup is a sporadic disease, in which respect it differs from pseudo-membranous croup, which is occasionally epidemic. Rilliet and Barthez, however, state that "it is incontestable that it may prevail epidemically," but this opinion is based not on their own observations, but on those of Jurine, of Geneva, who describes an epidemic which raged in that city in 1808. My own opinion is that the so-called epidemics of this disease depend upon certain conditions of the atmosphere exciting the morbid state to an unusual degree, and that it is never epidemic like whooping cough or even pseudo-membranous croup. Again, the disease is hereditary in certain families, and almost every physician of experience can call to mind families in which it has prevailed generation after generation. Dr. J. F. Meigs, of Philadelphia, remarks, "I am acquainted with one family in this city in which the children for three generations were extremely liable to it; with another, in which the grandmother and grand-children were frequently attacked; and with a third in which the father and children showed the same predisposition in the most marked manner."

This disease occurs alike in the robust and in the weak, and many children are predisposed to it when laboring under any digestive derangement. The most important exciting cause is exposure to cold, either sudden transitions from heat to cold or exposure in the open air. Narrowness of the rima glottidis is at times a predisposing cause, and nervous children are at all times most likely to be attacked. Dr. J. Lewis Smith has observed that this disease is not uncommon at the commencement of measles, and Dr. Condie notes that after an attack has once happened, the occurrence of any sudden or violent mental emotion is liable to excite a paroxysm.

It is characteristic of spasmodic croup to attack suddenly and without warning; as old Dr. Meigs quaintly puts it, "there is often not the least reason to suppose the child sick until the moment of explosion of the attack, an attack which in many examples is more violent in the first moment of its existence than in any subsquent time." In a considerable number of cases the attack is preceded by a paroxysm of teething fever, and so close is the connection between the fever of dentition and spasmodic croup that Dr. Copland affirms, "I have scarcely ever seen a well-defined case unconnected with dentition." In a much larger number of cases there is slight coryza with hoarse cough. Now, hoarseness excites little attention in adults, as in acute cases is does not usually indicate any special degree of danger, yet the contrary is the case with children, as with them *hoarseness always indicates danger*, and it should never be neglected. The first paroxysm generally takes place in the night during the first sleep, between ten o'clock and midnight. Out of a hundred cases ninety-five will occur in the night, and the remaining five in the afternoon, and three-fourths of the night cases will take place before midnight, and the remaining fourth after that hour. With or without, then, any premonitory symptoms, the child is attacked by a dry, ringing clangorous cough, which has been compared to the notes of a trumpet mingled with the rasping of a large saw, but which, as Professor Wood remarks, "is, in fact, comparable to nothing else in nature, and to be appreciated only by being heard." This sonorous and barking cough is accompanied by prolonged inspiration, by a shrill and rasping sound, and by rapid and irregular respiration. At times the breathing is so very irregular that suffocation appears to be impending, and the child tosses about in its bed as if fighting for air. The characteristic cough is, according to Wood, occasioned, in all probability, by a certain spasmodic rigidity of the vocal cords, giving an almost metallic tension to the sides of the rima glottidis. The voice is hoarse and rough, though rarely suppressed, and then only for a brief period— this is one of the most salient points of difference between the disease under consideration and pseudo-membranous croup. There is but little real pain in the larynx or trachea, but a feeling of constriction which seems to be still less endurable than pain. The little patient may endure the attack for a little time with considerable fortitude, but soon he sits up in bed gasping for breath, or lies on his back with his neck stretched to the utmost, while the throat is grasped by the hands as if to remove some obstacle to respiration. If able to speak, he complains of pain and tightness at the throat, while the face has an anxious and troubled expression. He becomes greatly agitated, cries violently between the fits of coughing, and begs piteously for help. When the paroxysm first comes on the face is flushed, the skin warm, and the pulse strong and frequent, but as the attack becomes more intense the face becomes of a livid hue, while the extremities are cool and the pulse frequent, feeble and fluttering. Copland says that "there is little or no increase of animal heat or fever," but fever was present in the vast majority of my cases, and there is a striking consensus of opinion on this point. After lasting from twenty minutes to an hour, or even two hours or more, the breathing becomes easier, the cough less frequent and less clangorous, and the sawing sound is only heard when the little patient cries. Often as soon as relief is obtained, the child falls into a sweet sleep. In the morning he seems nearly well, having only an occasional croupy cough, with hoarseness of the voice and redness of the fauces. At times this croupous cough continues for several days, gradually getting milder and less frequent, till at last it ceases entirely. When the attack occurs early in the night, it is likely to be followed by a second milder attack towards morning ; as the second evening approaches, the patient is the subject of a similar paroxysm of varying severity. Professor Wood says that the symptoms are often more violent than at first; Dr. J. F. Meigs says that, as a general rule, the first attack is the most severe. So far as my observation extends, I have noted that while the paroxysm is more prolonged and more exhausting, the symptoms appear to be less severe.

This may be taken as a fair account of a case of moderate severity, but at times the laryngeal inflammation is more intense and apparently involves a larger extent of the mucous membrane. The cough is hoarser and more frequent, the respiration more difficult, the fever more pronounced, and this state is developed, as a general rule, earlier in the night than the milder form of the disease. As the night advances all the symptoms intensify, till actual suffocation is threatened. Towards morning amelioration takes place, the fever declines, the breathing becomes easier, the cough looser, the stridulous sound less marked. But as the next evening approaches, all the symptoms reappear, to be followed by another remission during the day, which, in its turn, is followed by another night attack, and so on, for several days. I have attended a number of cases in which the daily remission hardly existed, for the disease continued day and night, for three or four days. If no fortunate change is brought about by treatment, the breathing becomes more and more difficult, the cough rarer and more feeble, and is finally suppressed altogether, the voice is hardly audible, the pulse is small and rapid, the face pale and cool, with pinched and contrasted features. Finally the child becomes comatose, and death takes place from asphyxia, often with general convulsions. In favorable cases the fever declines, the voice becomes stronger, the cough looser, the stridor diminishes, and the patient rapidly enters on convalescence. These severe cases last longer than those of the milder type. A severe case will keep the patient in real danger for two, three, or even four days, while the milder cases subside after forty-eight hours, the patient rarely being in real danger. In severe cases the disease is often followed by hoarse cough with husky breathing, and this state is sometimes difficult of cure. The disease is very likely to return, and paroxysms more frequently come on at intervals of from six to twelve months than in a shorter period, though I have attended patients who have had five or six attacks in a year.

Auscultation of the larynx should never be neglected, and for this purpose I now prefer the single stethoscope to the double one. On auscultation it will be found that the respiration is dry and wheezing, with a hissing, sonorous sound, as if the larynx were narrower than usual, and had rigid and unyielding walls. It is well to bear in mind the remark of M. Trousseau, that the hoarse-sounding, croupal cough is not a sign of exudation in the larynx, but rather of its absence.

Spasmodic croup is rarely fatal, so that we are not so conversant with its morbid anatomy as we are with that of the more formidable pseudo-membranous croup, and many of the post-mortem changes are wholly inadequate to account for the fatal result. The pathological state present is slight inflammatory hyperæmia, with perhaps increased activity of the mucous follicles. Very severe cases have redness of the larynx, extending to the trachea, or even to the bronchi, and this redness may either be continuous or in patches. There is usually a slight swelling of the sub-mucous tissue, with viscid and adherent mucus if death has taken place at an early stage of the disease, or with more abundant and purulent mucus if the disease proved fatal at a later period. Many have supposed that death often takes place from a literal occlusion of the larynx; but this is very seldom the case, for, in a vast majority of cases the larynx is sufficiently open for the purpose of respiration, and we conclude that many patients are asphyxiated by spasmodic contraction of the laryngeal muscles. " In some rare instances, no signs of disease are discovered in the mucous membrane, and the patient has probably died of spasm, consequent upon high vascular irritation or congestion, the marks of which disappear with life." (Wood.)

Almost the only disease with which spasmodic croup is likely to be confounded is pseudo-membranous croup, and the diagnosis will be carefully considered in the next chapter. It may also be mistaken for spasm of the glottis; the diagnosis will be found in the chapter on that disease.

Spasmodic croup is very rarely a fatal disease under homœopathic treatment. Still, cases which have been badly managed under other practitioners, may die under the care of an homœopathic physician; and the worst kind of mismanagement consists in the administration of violent emetics for the purpose of "clearing out the phlegm and breaking up the spasm," according to the wont of our allopathic step-brethren. I am satisfied that the irritant action of these emetics is one of the principal means whereby mild cases of croup are converted into severe ones. Very shrewd is the remark of Dr. J. Lewis Smith, "While a favorable opinion in reference to the result may ordinarily be expressed, the physician should not forget the fact that death may occur." One would say that a certain amount of danger is present when the disease lasts longer than forty-eight hours, and that the danger increases with the prolongation of the disease. When the paroxysms diminish in intensity, when the fever declines and the cough becomes moist, a favorable termination may confidently be expected. On the other hand, an extremely small and rapid pulse is an unfavorable sign, especially when met with in conjunction with coolness of the extremities; a marked intensity of the stridulous sound, especially in the expiration; suppression of the voice and extreme dyspnœa; paleness of the face and diminution of strength would form an extremely ominous group of symptoms. Should convulsions supervene in addition, there would remain no ground for hope.

Dr. Duncan, of Chicago, accurately remarks that "the treatment of spasmodic croup has been so mixed up with that of membranous croup and laryngitis (simplex) that the literature is very unsatisfactory," and he correctly adds, "the advice of Benninghausen to give Aconite, Spongia and Hepar is about as good as any routine treatment." Aconite is one of the leading remedies in the treatment of both spasmodic and pseudo-membranous croup, though Dr. Alphonse Teste says that it "is indicated in croup in the rare cases of violent fever in the beginning; from the moment that the febrile symptoms diminish a little, or when, after one or two doses, it seems to produce no effect, it should be discontinued, and that finally, under penalty of losing precious time, when often the minutes are to be counted." On the other hand, Croseric says "the first medicine to employ when the croup declares itself is Aconite," and Duncan remarks that "Aconite is often the only remedy needed, as it corresponds to the spasm, the restlessness, the anxiety and the fever that arises." Aconite is indicated when the patient is attacked in the evening after sleeping, though he has been restless and feverish before going to bed. The patient has great nervous and vascular excitement, with restless tossing about in the bed. On attempting to swallow he complains of pain in the throat, and this is aggravated during deglutition, though it is never really absent. The cough is dry, hacking and frequent, and it follows every expiratory effort, but is absent during inspiration. Sometimes the patient wants to cough, but restrains himself on account of the pain. The cough, as well as the stridulous sound, is distinctly paroxysmal, and it is characteristic that these are heard only during inspiration. The pulse is accelerated, the skin dry and hot, and the patient drinks with avidity. "The children attacked with this form of croup are the nervo-sanguine or nervo-bilious, *i. e.*, the nervous active, while membranous croup attacks lymphatic children as a rule. The more nervous the child the longer the spasm continues" (Duncan). In mild cases the dilutions will suffice, but when the case is threatening I put two or three drops of the tincture of the root in a tumbler of water, and give a teaspoonful every half hour, or even every fifteen minutes.

By almost universal consent Spongia occupies the next place to Aconite in the treatment of croup. Dr. Hughes says that "the two leading remedies in croup are Aconite and Spongia," while Professor Hempel says still more emphatically that "in this disease we have found that if Aconite and Spongia leave us in the lurch, the chances of recovery are very slight indeed." On the other hand, Dr. Teste says, " The good effects of Spongia are incontestable, but they have been exaggerated; Spongia belongs to the first period. I do not use it." Hahnemann says of Spongia: " Its most remarkable therapeutic virtue is to cure croup; among other symptoms it is indicated in this disease by difficulty in breathing, as though a plug had lodged in the throat, and as though the larynx were so constricted that breath cannot pass through it." Hempel recommends it to be given if Aconite produces a profuse, warm perspiration, and the spasmodic breathing still continues. In the croup of Spongia the larynx is painful, as if pressed, with a burning and constrictive sensation in the organ; respiration is difficult, as if a plug were in the throat; it is wheezing, hissing and sawing, and at intervals there are suffocative fits, during which the child is unable to breathe except with the head bent backwards ; the cough is hollow and barking, with difficult expectoration of scanty mucus; coughing causes pain in the trachea and lungs. Dr. Paine, of Bath, Maine, observes : " It seems that Spongia nearly covers the same symptoms as Aconite, with this difference and addition : in Spongia croup the stridulous respiratory sound is always during *inspiration* and the cough less constant, and excited only by the inspiratory act ; and the cough and sibilant respiratory sound are not so constantly concurrent as the Aconite croup. There is also, in Spongia croup, fluent coryza, and sometimes sneezing, with saliva dribbling from the mouth, which we do not see in Aconite croup." I remember hearing Dr. Constantine Hering remark that Spongia has aggravation in the evening, while Hepar has aggravation in the morning. The cough of Spongia is piping, crowing, and of a very dry sound, with rough, crowing cry, and sensitiveness to the touch ; while the cough of Hepar is deep, rough and barking ; hoarseness or aphonia, with slight suffocative spasms ; respiration not without rattling of mucus. The cough of Spongia is worse when sitting erect and better in the horizontal position, while the cough of Hepar is excited by lying and is aggravated by lying with the head low, and is better with the head high. The cough of Spongia is improved by eating and drinking, while the cough of Hepar is excited by cold diet. Hempel recommends the use of the tincture. I have always used the 2d or 3d decimal dilutions, prepared and administered in the same manner as Aconite.

In routine practice—and some of Hahnemann's followers adhere to routine as abjectly as did the Esculapians whom Hahnemann scourged—it is customary to give first Aconite, then Spongia; next, "if that don't do," Hepar. Even Hartmann, who is not usually a *routinier*, recommends Spongia to be given after Aconite, adding, "in twenty-four hours the danger is generally over. If, after this lapse of time, the cough should still have the peculiar croup sound, the breathing should still be hissing, or if there should still be danger of suffocation, Hepar is then to be employed." The dry, harsh, deep and hollow cough of Hepar is apparently caused by tickling in the larynx or scraping in the trachea, and is increased unto vomiting by a deep inspiration ; there is a constant mucous rattling from which the patient vainly endeavors to obtain relief by expectoration ; there is pressing in the throat, with a constrictive feeling as if he would be suffocated. The respiration is exceedingly quick and laborious, and the voice is hoarse and weak. The skin is dry and burning, and the patient is restless and inclined to weep. Dr. Hughes does not assign Hepar a prominent place in the treatment of croup, merely stating that it is useful in restoring the laryngeal membrane to its normal condition when the croup is hoarsely mucous ; and Dr. Duncan considers that the remedy is indicated "after the spasm is relieved, and there is a loose, hoarse cough, worse towards evening, with a little fever, due to the obstruction of the mucus." Dr. Edmonds gives this remedy " for the remnant of symptomatic *debris* that may be found on hand towards the conclusion of the case, mainly in the shape of a cough, which seems rather irritative than inflammatory :" Dissolve a grain of the 4th or 5th decimal trituration in half a tumblerful of water, and give a teaspoonful every half hour or oftener.

Phosphorus is one of the remedies rarely given at first, but held in reserve, as it were, in case the other remedies should fail. Kreussler says, " If these three remedies (Aconite, Spongia and Hepar) should remain ineffectual, or should only effect a partial cure, Phosphorus is still left us." In like manner, Laurie advises Phosphorus " in cases where Hepar may fail to remove the symptoms we have enumerated under that remedy ; or where Aconite and Spongia, as well as Hepar have been merely productive of temporary benefit." The cough of Phosphorus is dry and tickling, but not very harsh sounding, with hoarseness and pain in the chest as if excoriated, and a continual irritation in the larynx and trachea, with shortness of breath ; or expectoration of mucus, with hollow cough. Phosphorus closely resembles Hepar, but it differs materially from Spongia. In Phosphorus the voice is trembling and hissing, while in Spongia the voice is interrupted. The respiration of Phosphorus is generally quick, while in Spongia it is predominantly slow. In Phosphorus the expectoration is most constant morning and during the day, while in Spongia the cough is generally dry; expectoration not constant ; is loosened in the morning and swallowed. Rückert remarks that Phosphorus is sometimes given with benefit when the improvement seemed to stop ; it did not, however, accelerate the cure.

Hartmann recommends it for obstinate hoarseness with slight catarrhal croup remaining after the disease is cured, and he also gives it for the tendency to relapse. Some twenty years ago I attended a case of croup, and the child got well, except a hoarseness which remained and excited my suspicion. But the parents insisted that the child was doing well. Next day another attack of croup destroyed the child's life. Since then I have never failed to use the Phosphorus under similar circumstances." I have found Phosphorus of great use for the weakness remaining after the attack, and when given at long intervals, and in the 12th to the 30th dilution, it often removes the predisposition to the disease. Of late years, I have found Sanguinaria still more effective for the same purpose. In acute cases I use the 5th or 6th decimal dilutions, but the removal of the predisposition requires higher potencies.

Our practitioners may thank Dr. Duncan, of Chicago, for recalling their attention to Lobelia inflata as a remedy for spasmodic croup, a remedy which had almost passed out of our minds, but which has done excellent service in the past. "Lobelia cases resemble those of Aconite with this difference: there is more dyspnœa and the spasm affects the œsophageal muscles, impeding deglutition as well as respiration. Older children will describe a sensation of a lump in the throat (*Ign*), but the constant ringing cough, stridulous breathing, and great anguish and fear of suffocation, distinguish the case from *Ignatia* or *Aconite*" (Duncan).

In pressing cases it would be well to consult the therapeutics of pseudo-membranous croup in the next chapter, especially the remarks on Sanguinaria, which for many years has been my sheet-anchor in both forms of croup.

The little patient should be kept as tranquil as possible, and this is just as necessary during the interval as it is during the attack. The atmosphere of the sick-room should be pure and of equable temperature, but draughts should be carefully avoided; I have known a number of serious relapses from neglect of this self-evident precaution. I have seen decided benefit from charging the atmosphere of the room with the vapor of warm water. The dress worn during the illness should be loose and easy, and a woollen wrapper should be worn in addition to the usual night-dress. If possible, the child should be kept in bed during the entire time of the acute attack, but a patient would be safer up and dressed than sometimes in bed and sometimes out of it —all in the night-dress. Condie attaches a good deal of importance to supporting the child in an erect posture during the paroxysms, and I have found that respiration is much easier when this simple recommendation is followed.

Almost all medical writers recommend a warm bath, say of the temperature of 100°, as soon as possible after the commencement of the attack, with the view of relieving the spasmodic action of the laryngeal muscles. 'I was in the habit of using this in former years, but of late I have discontinued it, as I found the reaction from it exceedingly injurious, and I now look upon the warm bath as a decided injury to the case. I have, however, seen benefit from the application of a sponge soaked in hot water to the region of the larynx and trachea, and repeated say every fifteen or twenty minutes. Really this is a counter-irritant, acting by revulsion from the larynx, but in a great majority of cases the dyspnœa, cough and hoarseness diminish at once, and I have never seen any bad result from it.

The food should be light and easy of digestion, and should consist of bread, rice, arrow-root and the various preparations of milk. I have seen great good follow the use of well-made beef-tea, a tablespoonful every two or three hours.

Condie observes that "when the paroxysm is very violent and long-continued, and there is danger of the occurrence of asphyxia unless immediate relief is obtained, the operation of tracheotomy should be performed without delay. But under enlightened homœopathic treatment, this must very rarely be necessary, save when the type of disease changes and the malady becomes pseudo-membranous croup.

How can we prevent spasmodic croup? I have succeeded in a great many cases in which the predisposition continued long after the completion of the first dentition by the persistent use of Phosphorus, say twice a week, on going to bed. In addition, as a matter of course, I attended to the usual prophylactic treatment of children subject to this disease. The dress should be warm and comfortable, for the custom of exposing the whole of the neck and a good part of the chest, as well as the upper limbs and the lower ones from the knee to the ankle, is one of the worst follies ever perpetrated in the sacred name of Fashion. Children subject to croup should wear woollen underclothing, light but warm, from head to foot, and the night-dress should be made in the same fashion. Dr. Eberle mentions, as showing the influence of dress, that during a practice of six years among the Germans, who keep the necks and chests of their children carefully covered, he met with but one case of this disease; and it is a comparatively rare disease in Montreal, where children are as rationally dressed as adults. Exercise in the open air should be taken whenever the weather permits, and the presence of snow should not be a barrier to a walk if the feet are properly shod. For many years I have followed the excellent advice of Dr. J. F. Meigs: "When the liability to the disease continues after the completion of the first dentition, I have found the daily use of the cold bath, in connection always with warm clothing, most useful in preventing the attacks. The bath must be commenced with in the Summer, and persevered in through the following Winter. The water, after the cold weather begins, should be drawn in the evening, allowed to stand all night in a room in which there is a fire through the day, and made use of on the following day. Prepared in this way I have found the water in the morning at a temperature of between 50° and 60° F. The child ought to be kept in the water only half a minute or a minute, then well rubbed and dressed immediately."

Aphorisms.

1. Spasmodic croup is a combination of catarrhal laryngitis and violent spasm of the interior muscles of the larynx.

2. Boys are more frequently affected than girls, and it is most common during the first dentition.

3. Spasmodic croup is never epidemic in the proper sense of the word, though it is sometimes hereditary.

4. Spasmodic croup is rarely a fatal disease, still the physician should not forget the fact that death may occur.

5. Danger is present when the disease lasts longer than forty-eight hours, and the danger increases with the prolongation of the attack.

6. The homœopathic remedies are Aconite, Spongia, Hepar, Phosphorus, Lobelia and Sanguinaria.

7. The best prophylactics of spasmodic croup are warm clothing, judicious exercise, regulated bathing, and the persistent administration of Sanguinaria and Phosphorus.

CHAPTER VIII.

Pseudo-Membranous Croup.

This is one of the most dreaded, and, till the advent of the homœopathic healing art, one of the most fatal of all the diseases of childhood ; and even with all the resources of the Similia, the thoroughly educated physician feels some little trepidation when he finds himself face to face with a well-marked case of this disease. Here, as in many other instances, immense advantage is derived from a thorough knowledge of the pathology and pathological anatomy of the disease, and when to this is joined a thorough knowledge of our Materia Medica, the homœopathic physician is better armed than the practitioner of any other school whatever. The contemptuous ignorance of pathology and pathological anatomy is thus keenly reproved by one of the most brilliant writers of our school: "It is because the *blind application* of our therapeutic law so often helps us when we grope vainly for the pathology, that we are led into a contempt for *pathology and such allopathic studies*. As healers we might be content with our therapeutic law; but as physicians we aver it is our duty to our profession to develop its every branch. To-day we often do not know what we have cured; and while *knowing* the Materia Medica will increase our capabilities for curing, it will not enlighten us in diagnosis and pathology."

Like spasmodic croup, this disease has had a multitude of names, many of which are mere misnomers. Guersant calls it 'pseudo-membranous pharyngo-laryngitis;' Rilliet and Barthez style it pseudo-membranous laryngitis, in which they are followed by Dr. J. Lewis Smith of New York; while other French writers persist in calling it laryngeal diphtheria; Bæhr of Hanover calls it 'Laryngo-tracheitis Crouposa,' an uncouth name, but anatomically and pathologically correct. Fletcher of Edinburgh—most homœopathic of all allopathic pathologists—selects this disease as a specimen of the preposterous names with which nosologists have labelled disease. "Croup, which has successively borne the names of suffocatio stridula (Horne), catarrhus suffocativus (Hillary), cynanche stridula (Crawford, Wedderburn), angina inflammatose infantilis, angina epidemica (A. Miller), angina polyposa, angina suffocativa (Baird), asthma infantilis (Millar and Bush), morbus strangulosus, plastic inflammation of the air-passages (Laennec), diphtheritis (Bretonneau), has lately been dignified with the name of dento-frangibalus-broncho-laryngo-tracheitis-mixo-pio-meningitis, and this probably is but a single specimen of what we must expect if this mania be not resolutely checked." While some of these names are simply ludicrous, others are really pernicious nonsense with a distinct tendency to mislead the anxious physician. For example, Dr. Condie, one of the best writers on children's diseases on this continent, heads one chapter "Tracheitis-Croup;" and two of the most recent writers on the subject—Sir George Duncan Gibb and Professor Aitken—adopt Cullen's erroneous name of "cynanche trachealis," and Sir Thomas Watson styles it "cynanche-trachealis-tracheitis-croup," and adds: "The essence of this complaint is violent inflammation, affecting the mucous membrane of that portion of the air-passages which lies between the laryngeal cartilages and the primary bronchi—in one word, of the trachea or *windpipe*. This is the genuine seat of the disease; but the inflammation sometimes ascends into the larynx; and not unfrequently it dives into the bronchi and into their ramifications." Now, pseudo-membranous croup—which I conceive to be the most appropriate name—in a large proportion of cases, commences in the larynx and extends *downwards*, and it is compatively seldom that it commences in the trachea and extends *upwards*, though in some cases a pseudo-membranous inflammation may extend from the bronchi to the trachea.

Pseudo-membranous croup is an inflammation of the epiglottis, glottis and larynx. frequently extending to the trachea, and occasionally reaching to the larger bronchial tubes, and this inflammation is accompanied by the exudation of a yellowish-white fibrinous material upon the mucous membrane of the affected parts; the fauces and tonsils frequently exhibit more or less of the inflammation with its accompanying exudation. The disease, then, is, in the words of Da Costa, "not only inflammation, but inflammation which results in the formation of a false membrane," and it must be specially noted that these membranes produce no loss of substance, and that they leave behind them no cicatrices. As a result of these morbid changes the breathing is difficult, loud and accelerated, with shrill or wheezing inspiratory murmur; the voice is at first hoarse and rough, but towards the close whispering or extinct; spasm of the interior muscles of the larynx is almost invariably present, and towards the close of the disease fragments of false membrane are sometimes expectorated or vomited. Fever is an almost invariable concomitant.

M. Littré, who in addition to his gigantic labors as a exicographer, was a medical writer of great merit, discusses the question whether or not Hippocrates was acquainted with croup, but he does not give any decided opinion on the matter. The following passage, however, certainly seems to apply to this disease: "Angina Gravissima quidem est, et celerrimè interimit, quæ neque in faucibus neque in cervice quicquam conspicuum facit, plurimum verò dolorem exhibet, et difficultatem spirandi, quæ erectâ cervice obitur, inducit. Hæc enim eodem etiam die, et secundo, et tertio, et quarto strangulat." Dr. Francis Adams, the learned commentator on Hippocrates and Paulus Aegineta, considers that there can be no doubt that the ancients were well acquainted with that species of cynanche in which the disease spreads down to the windpipe. Few of us, however, would agree with Dr. Adams when he says, "It may reasonably be doubted whether they (the ancients) were not fully as well acquainted with diseases of the fauces and windpipe as the moderns are." Baillon (Paris, 1576) was the first writer of modern times to describe this disease, "Chirurgus affirmavit se secuisse cadaver pueri ista difficili respiratione et morbo (ut dixi)incognito sublati; inventa est pituita lenta, contumax, quæ instar membranæ cujusdam arteria aspera erat obtenta, ut non esset liber exitus et introitus spiritui externo, sic suffocatio repentina." According to Fredrich, Baillon was the first who mentions having dissected a patient who had died of croup. Boerhaave and Willis describe morbid states strikingly like croup, and the "suffocative catarrh" of Ettmüller is clearly croup under another name. Dr. Blair of Cupar Angus, in Scotland, first described the disease by its present name, in the year 1713. "The tussis convulsiva or chink-cough, is also some years epidemical, and becomes universal among children; as is a certain distemper with us called the croops, with this variety, that whereas the chink-cough increases gradually, is of long continuance, seizes in paroxysms, and the patient is well in the interval; this convulsion of the larinx, as it begins so it continues, so violently that unless the child is relieved in a few hours 'tis carried off within twenty-four, or at most forty-eight hours. When they are seized they have a terrible snorting at the nose and squeaking in the throat, without the least minute of free breathing, and that of a sudden; when perhaps the child was but a little time before healthful and well. The most immediate cure is instant bleeding at the jugular, either by the lancet or leeches; when the most urgent symptoms are gone, then emetics or the like are administered at discretion." Two French writers, Molloi and Malain, described the disease in 1743 and 1745 respectively; in 1749, Ghizi, of Cremona, gave a good account of it under the name of *angina strepitosa*, and in the same year it was described by Dr. Starr, of Liskeard, in Cornwall. In 1755, Dr. Richard Russell, of London, described the disease as observed by him in connection with the epidemic sore-throat then raging, and he points out that "it is most apt to seize children from two years old to eight or ten, but chiefly the younger sort"—most probably laryngeal diphtheria. Dr. Francis Home's essay, entitled "An Enquiry into the Nature, Cause and Cure of Croup," appeared in 1765, a carefully written account of true croup as observed in Edinburgh and the neighboring towns at a time when the disease was not complicated by epidemic affections of the fauces. Home regarded the disease as an acute inflammation of the larynx and trachea, and his descriptions, with those of his Swedish contemporaries, Halen and Wahlbom, gave croup a definite place among diseases. Millar, who practiced in Scotland at a later date than Home, published his "Observations on the Asthma and Hooping Cough," in 1796, and he gives greater prominence to the spasmodic element in the disease than any previous writer—indeed he seems to have spasmodic croup in his mind's eye more than the true croup of Home.

Some of these writers, notably Ghizi of Cremona, and Starr of Liskeard, were describing what we would call diphtheritic croup, and Dr. Richard Russell was clearly describing an epidemic angina which often extended to the larynx and trachea, and which was entirely distinct from the sporadic and purely inflammatory disease so ably described by Home. The latter writer thus contrasts the two diseases: "The two very different situations of the suffocatio stridula ; the former more inflammatory and less dangerous ; the latter less inflammatory and highly dangerous ; in the former the pulse is generally strong, the face red, drought great, and they agree with evacuations ; in the latter the pulse is very quick and soft, great weakness, tongue moist, less drought, great anxiety, and evacuations hasten death." In spite of these clear diagnostic differences, many epidemics of angina maligna—which would now be styled diphtheria—in the eighteenth century were called croup, just as not a few cases of laryngeal diphtheria are included in the accounts of croup written near our own day.

In the first year of the nineteenth century, Cheyne published in his "Essays on the Diseases of Children," a treatise on cynanche trachealis in which he maintained the views of Home with great ability, and many years later the same learned physician wrote the article on croup in the *Cyclopædia of Practical Medicine*, in the hands of so many practitioners on this continent. From 1805 to 1807, a great epidemic of croup, so-called, swept over the western part of the continent of Europe, and among its most illustrious victims was the Crown-Prince of Holland, nephew of Napoleon the Great and brother of Napoleon III. His uncle, who was tenderly attached to the lad, ordered the institution of the famous *Concours* on croup, and of the 83 essays sent in, though many describe diphtheritic croup, the writers who carried off the principal prizes, Jurine of Geneva and Alhers of Bremen, unquestionably describe an independent disease, inflammatory in its nature, uncomplicated by malignant angina or any epidemic influence whatever. Still later in point of time, we had a controversy, hardly terminated, between the observers who contended that there was but one form of croup, and that other body of practitioners who drew a sharp line of demarcation between pseudo-membranous croup and spasmodic croup.

In the last chapter the writer remarked upon the fact that spasmodic croup is much more frequent than the pseudo-membranous variety, and it was stated that while spasmodic croup is a disease of very young children, pseudo membranous croup generally affects those of more mature years. Cullen remarks, "This disease seldom attacks infants till after they have been weaned. After this period, the younger they are, the more they are liable to this disease. The frequency of it becomes less as children become more advanced; and there are no instances of children above twelve years being affected with it." The last assertion is unquestionably an error, for it has been often seen in its most formidable form in children at the breast, and adults have died from it. Bæhr of Hanover questions whether adults have ever died of genuine croup, and he remarks that it occurs even less frequently before the second than after the seventh year. This last remark is not in harmony with the experience of any other writer, and is contradicted, moreover, by the Vienna statistics, noted for their accuracy. "Among 501 deaths from croup in Vienna, during 1868, 92 were in the first year (30 were 12 months old, and 12 were 11 months), 128 in the second, 87 in the third, 71 in the fourth, 50 in the fifth, 34 in the sixth, 17 in the seventh, 7 in the eighth, 6 in the ninth, 2 in the tenth and eleventh respectively, 3 in the twelfth, 1 in the thirteenth, and 1 in the sixty-second." (Glatter.) Dr. Condie says that in Philadelphia, during the ten years preceding 1845, 319 deaths were reported from croup in infants under one year; 238 in those between one and two years; 475 in those between two and five years; 112 in those between five and ten years; and 6 in children over ten years. "Of 2,136 fatal cases reported in this city (Philadelphia) during the seven years from 1862-68, 301 were under 1 year of age; 571 between 1 and 2 years; 951 between 2 and 5 years; or, 1,522 between 1 and 5 years; and 236 between 5 and 10 years; leaving but 77 cases as occurring after the latter period of life (Meigs and Pepper). The same writers add that of the 35 cases that they have seen,

28 occurred between 2 and 7 years of age, while of the remaining 7, 1 occurred at the age of 18 months; 1 at that of 19 months; 1 at 7½ years; 2 at 11 years, and 1 each at 11½, and 12½ years. In 1870, 4,302 children died of this disease in England, of which 3,663 were under five years of age, and the largest number of deaths was amongst those in their second year. Instances are reported in which the disease occurred at a very early age. Morley and Cheyne speak of having seen it in infants of less than three months, and Bouchut has seen it in one only eight days old. Dr. Home remarks that the younger children are when weaned, the more liable are they, *cæteris paribus*, to this malady, and this has been confirmed by Cheyne and other excellent observers.

In contrast with whooping-cough which principally affects female children, pseudo-membranous croup is much more frequent among males than females, though Meigs and Pepper state that "sex cannot be said to exercise any decided influence upon the frequency of this disease," and in support of this statement, they point out that of the above-mentioned 2,136 cases, 1,115 occurred in males and 1,021 in females. Dr. Squire is almost of the same opinion as the distinguished Philadelphia writers. He says: " More boys than girls are born in a proportion somewhat greater than one in every fifty children, or, to give the result of a very extended examination, there are 511.75 males and 488.25 females in every 1,000 births ; it appears that of this number 83.71 males and 65.74 females die within the first year, after which the death ratio of the two sexes for the next ten years is nearly equal ; still there are a larger number of males than of females living at that period, and the deaths of females from all causes are to those of males as 87 to 100 in the first five years, or as 88 to 100 in the first ten years. Now the deaths from croup are so nearly in this proportion, and of late years have so often shown a difference so much less than this, that a doubt might be entertained as to whether any difference in the liability of the sexes really existed. A comparison between the deaths from all causes of each sex for each year with the deaths from croup at each year, sex with sex, shows a difference of excess on the side of the males so constant that it is rare to meet with an exception, but at the same time so slight that it can only be considered a characteristic of the disease in the aggregate, corresponding with the results of pneumonia and tubercular meningitis rather than with the more characteric zymotic diseases, and contrasting with those of diphtheria and whooping-cough, where the excess of deaths is greatly on the side of the females."

On the other side of the question, without giving figures, Felix von Niemeyer states that " boys are more subject to it than girls," and Bæhr thinks that from 60 to 70 per cent. of all cases are boys. Of 429 deaths from this disease in Massachusetts during 1852, 243 were in boys and 178 in girls. Of Steiner's 101 cases, 77 were boys and only 24 girls ; of 30 cases reported by Trousseau 22 were males and 8 females. Of Bohn's 70 cases 43 were boys and 27 were girls, while of Jauseconich's 22 cases 17 were boys and 5 were girls. Ruehle giving the proportion of boys to girls as 3 to 2 ; and the deaths from croup in the London Hospitals during the year 1840 were *three* in the male sex to *one* in the female. Glatter makes the curious observation that among the Christian population of Vienna the mortality from croup is 2.6 per cent., while among the Israelitish it is 4.2 per cent. and Steiner confirms this from his own experience in Prague. The writer's own experience is, that not only is the disease more frequent in males than in females but it is, at the same time, more severe and more fatal, so that a little girl's chances of life are much brighter when attacked with pseudo-membranous croup than are the chances of a little boy.

Almost all observers are agreed that pseudo-membranous croup is largely a malady of children of sanguineous temperament, plump and of ruddy complexion, and apparently in the enjoyment of excellent health. Drs. Meigs and Pepper remark of their 35 cases that 26 occurred in healthy, vigorous children, while the remaining 9 occurred in children who though neither very weak nor very sickly, presented a rather delicate appearance. Steiner, in one article on croup, says "that true croup appears by preference in strong children, well nourished and previously healthy, and in another paper, of later date, he says that "strong, well-fed, hearty children are no more liable to croup than those who are feeble, delicate, or affected with other diseases." Felix von Niemeyer is the most distinguished writer who contradicts this view, so generally held : " It is an error to suppose that vigorous, full-blooded, blooming children are especially liable. On the contrary, tender, delicate ill-nourished offspring of tuberculous parentage, with pale skin and conspicuous veins (an ominous sign even for the laity), children with a tendency to moist eruptions, to enlarged lymphatics, or to acute hydrocephalus, suffer from croup with equal or even greater frequence than those who are more robust. It is our daily experience that, in the great mortality which desolates certain families, a portion of the members die of croup, and another of hydrocephalus, while in the survivors pulmonary tuberculosis develops later in life. It would appear that the croup not unfrequently begins very soon after the disappearance of a moist eruption on the head or face."

I consider that *season and temperature* exercise a much more powerful influence than *constitution and temperament* upon the development of true croup, for, like the spasmodic croup, it is, as Cullen remarks, 'often manifestly the result of cold applied to the body,' especially of sudden transitions from heat to cold. Professor Gölis of Vienna relates the case of a boy four years old, previously in perfect health, who, having gone out from an over-heated room into the open air during an extremely cold winter's day, was seized while walking with all the symptoms of the most violent croup, which proved fatal in fourteen hours. It is most common during cold, damp, changeable weather, and it often attacks children who live in over-heated rooms and who are taken into the open air without proper clothing, especially during keen east or north-east winds, or the prevalence of sudden changes of temperature. Croup is four times as frequent in the winter quarter, November, December and January, as in the summer quarter, June, July and August. Dr. J. Lewis Smith notes that it is common among the poor of New York, who live in close rooms, over-heated during the days and cool at night, and Drs. Meigs and Pepper formulate the opinion of the profession by saying: "*The mean monthly temperature and the mean monthly mortality from croup rise and fall together throughout the entire year.*"

Dr. Cheyne thought that the liability of children to this disease depended upon the narrowness of the chink of the glottis, and in support of this Drs. Evanson and Maunsell pointed out that there is "scarcely any perceptible difference between the aperture of the glottis of a child of three and one of twelve years of age; while, after puberty, that opening is suddenly enlarged, in the male, in proportion of ten to five, and in the female, of seven to five." Guibert thought that the straightness of the windpipe and particularly of the glottis, "rendered croup more frequent in infancy, but is difficult to see how it could have this effect." Finally, without going as far as Meigs and Pepper, who frankly admit that "in none of the cases that we have seen could the exciting cause be even suspected," I would remark that in very many cases the exciting cause is absolutely inscrutable.

Does true croup—pseudo-membranous croup—ever recur in the same patient? Dr. William Squire, of London, who has bestowed extraordinary pains on the etiology of this disease, is confident that it does recur: "Children who have suffered an attack are specially liable to a recurrence on exposure to any of these causes (exposure to cold, change of dress, etc.), and the recurrent attack is not always the least severe." Yet Dr. Squire is unquestionably describing true croup—not spasmodic croup—which was certainly in the minds of Evanson and Maunsell when they wrote "when a child has once been affected with croup, it must be considered liable to a recurrence of the disease at any period until the arrival of puberty." Steiner remarks that true croup, as a rule, occurs in the same child once only, though there are a few occasional instances of a second attack, and in his own experience of more than a hundred thousand cases of disease among children, he has never yet met with a single recurrence of true croup. Meigs and Pepper had two patients in whom second attacks occurred, a very large proportion of a total of 35 cases. The writer never met with a recurrence of pseudo-membranous croup, and stories such as that told us by Dewees, who claimed to have attended a lady of forty for five attacks of croup within six years, are just so many illustrations of the ignorance of the distinction between true croup and spasmodic croup. I am quite willing, however, to admit the correctness of Bæhr's observation, "if a child has been once attacked with croup, it retains an increased disposition to inflammatory affections of the larynx." But these "inflammatory affections" rarely assume the form of pseudo-membranous croup.

Steiner speaks of "a *certain hereditary and family disposition* to croupous inflammation in general, and to laryngeal croup in particular," and in discussing the subject he says, "I have quite recently become acquainted with two unfortunate families, in one of which all four, and in the other all three children died of membranous croup, within five years in the one case, and within four years in the other." I submit that unless Steiner can prove that the parents of these children had had membranous croup in infancy, it is clearly incorrect to speak of an *hereditary disposition*, though these are probably illustrations of a *family* predisposition, which is not so marked in true croup as in the spasmodic variety of the disease. It is but fair to admit that many writers of note vehemently deny that any such family predisposition exists. About twenty-two years ago I attended two families in whom the disease appeared in every child during the second or third year of life. The children were six in number, of whom I lost three; of these one was moribund when I first saw it.

Pseudo-membranous croup is, generally speaking, a sporadic disease, and though not so frequently seen as spasmodic croup, it is by no means such a rare occurrence as Cullen supposed it to be. Many writers contend that it is occasionally epidemic, and Dr. Churchill gives us a most formidable catalogue. "The principal epidemics of which we have authentic accounts are those of Paris in 1506 (Baillon); Cremona in 1747 (Ghizi); Cornwall in 1748 (Starr); Upsal, 1762 (Rosenstein); Frankfort in 1764 (Van Bergen); Sweden in 1768–72 (Wahlbom and Bœck); Wertheim in 1772 (Zobel); in Galicia, in 1778 (Hirshfeld); Clausthal in 1783 (Bœhmer); the United States in 1805 (Barker); Stuttgard, in 1807 (Autenrieth); Saxony, in 1807–8 (Albers); and again, in 1811 (Schundtmann); at Vienna, 1807–8 (Gölis); and in Maryland, in 1807 (Chatard)." Bouchut is quite certain that the disease is epidemic: "Croup is an epidemic disease. This characteristic is a difficult one to establish at Paris, where most of the cases are disseminated and lost as regards each medical man who is limited to a portion of the field of public health. There, there is no general epidemic; only partial epidemics are observed developed in a quarter, in a house, or in a hospital devoted to infants. Still more must these epidemics be declared very unfrequent, for only one has been observed at the hospital for children at Paris, and that not very well characterized. The epidemic character especially reveals itself in limited localities. It is impossible to mistake it when it is observed in a province and in districts where nothing is ignored, and where the ravages caused by this disease in the population can be closely followed." Dr. J. F. Meigs remarks: "When epidemic, it is very generally connected with angina, while the sporadic cases frequently begin in the larynx, and often run their course without implicating the pharynx. During the latter part of the year 1844, the whole of 1845, and a part of 1846, the disease prevailed extensively in this city, and was in many cases accompanied by the pharyngeal affection. During these years, and particularly in 1845, measles and scarlatina also prevailed to a great extent, especially the former." Steiner says that, "Primary croup occurs sometimes *sporadically*, sometimes, though less frequently, as an *epidemic*. When several children in a family, or a large number in a neighborhood, are affected with the disease, most of such instances belong generally to the epidemic form; but this distinction has not been sufficiently observed in the literature of croup to make it available for statistical purposes." "Not unfrequently," says von Niemeyer, "we observe its epidemic appearance. At such times many children are attacked, even in one small place, and often several children of the same family in quick succession, and by the most intense and pernicious form of the disease." Condie is as positive on this point as his distinguished townsman, J. F. Meigs: "Of the frequent prevalence of croup as an epidemic, Berge, Canstatt, Fleury, Valleix, Wunderlich, and others, furnish incontestable evidence. An epidemic of the disease is recorded as having extended over the greater portion of Central Europe during the period between 1805 and 1807, and one of more circumscribed limits by Ferrand in 'his Thesis on Membranous Angina, published in 1827, during which, in a district of very small extent, there occurred no less than sixty cases of croup, all terminating fatally." The writer has seen two epidemics of pseudo-membranous croup of limited extent, and in both, the disease commenced as an intense pharyngitis, which, however, was neither diphtheritic nor scarlatinous in its nature. One of these epidemics, which appeared in the year 1859 was the immediate forerunner of a very severe epidemic of diphtheria, of which but few cases affected the larynx.

Little has been written as to *endemics* of pseudo-membranous croup, but a number of years ago a series of facts was observed by the writer which leads him to believe that the disease may occasionally rage as an *endemic*. Briefly, the facts are as follows: A family, the children of which had not been subject to croup, moved into a house situated near a low and stagnant creek, and very soon several of the children had severe attacks of true croup. After a very sickly time, that family removed to another house and a second family took their place. But soon the second family was attacked by true croup in a very severe form, and they, too, concluded to change their quarters. Neither of these families had croup either before or after their residence in that house, the subsequent medical history of which I have been unable to trace. I have observed some other cases less marked than the above, and in my native city of Edinburgh the disease has been noted to prevail as an endemic in the Cowgate, which is a long and very squalid street, occupying the deepest part of a valley densely crowded by buildings of a very unhealthy nature. Sir Thomas Watson remarks: "Towns situated on the banks of rivers have more than the average share of it; and it has been observed to be particularly frequent among the children of washerwomen in such places, and thus evidently connected with exposure to moisture. In towns so situated, it has been known to prevail epidemically after an inundation."

Is pseudo-membranous croup contagious? This question is clearly wrapped up with the question of the identity or non-identity of pseudo-membranous croup and diphtheritic croup, which is fully discussed in the next chapter, but for the sake of completeness, the question of contagion will be considered here. Aitken avoids the question altogether: "While the annals of medicine are rich in descriptions of epidemic and endemic croup, opinions are very much divided as to the nature of the epidemic influence, and whether or not the disease is contagious or infectious." Bouchut maintains the contagious nature of croup, but his remarks eviedntly apply to diphtheritic croup: "Its contagious nature is far from being demonstrated; still this question must not be answered in the negative, for croup often follows pseudo-membranous angina. Now, the contagion of this latter disease has been demonstrated in the most positive manner by the observations of M. M. Bretonneau and Trousseau. It is, then, possible that croup, which by its nature very much resembles pseudo-membranous angina, may, like it, be transmitted by contagion. I say possible, for in the present state of science a more positive expression cannot be made use of. It is, consequently, proper to separate those children laboring under croup from other children whose health has not, as yet, experienced any attack." In a similar strain Churchill remarks: "Several authors, Wichmann, Brehmer, Field and others, maintain the contagiousness of croup; but this is denied by the majority of writers, at all events in the case of primary croup. Certain forms of diphtheritic inflammation of the fauces and pharynx are undoubtedly contagious; and as the inflammation and exudation sometimes spread to the larynx, constituting secondary croup, it may be so far regarded as sharing in the same mode of propagation." Steiner says, "Some authors regard ordinary inflammatory croup as infectious, but with this I do not agree, though there can be no doubt that the diphtheritic variety is eminently contagious;" and again, in a later paper, "Primary croup occurs sometimes *sporadically*, sometimes, though less frequently, as an epidemic. When several children in a family, or a large number in a neighborhood, are affected with the disease, most of such instances belong generally to the epidemic form; but this distinction has not been sufficiently observed in the literature of croup to make it available for statistical purposes." Bæhr tersely says "that croup is contagious is only believed by those who regard croup and diphtheria as identical," and Condie is equally explicit: "Under no circumstances do we believe croup to be contagious." Felix von Niemeyer observes: "In some croup-epidemics facts have been observed which make it somewhat probable that the disease may spread by contagion. It is questionable, however, whether there may not have been confusion with that highly-contagious malady, epidemic diphtheria; in these cases, as we shall hereafter demonstrate, the fact that secondary croup of the larynx often accompanies diphtheria of the fauces." Copland says, "it has most indubitably manifested this property (contagion) when it has prevailed epidemically, and when associated with cynanche maligna"—which, in our day, would be styled diphtheria. My old clinical teacher, George B. Wood, writes: "The disease has also been ascribed by some writers to epidemic and contagious influences. But, if we except the cases which are apt to occur during the prevalence of epidemic catarrh, it is only to the diphtheritic disease of Bretonneau that this remark is applicable. Original, uncomplicated croup is probably never either epidemic or contagious." Pseudo-membranous croup, then, may safely be set down as being non-contagious, while the reverse is the case with diphtheritic croup, and, personally, I agree with Dr. Squires: "Croup, indeed, seems to hold a place intermediate between the zymotic class and those of the respiratory organs."

As a general rule, true croup is a primary disease, but occasionally it is secondary; indeed, physicians are only now realizing the truth of Lefferts' remark, "*Unquestionably in the majority of cases of acute infectious disease the larynx is more or less implicated.*" West defines secondary croup to be "that form which occurs in the course of acute, infective or general constitutional diseases, pyæmic processes and other acute or chronic affections." Measles is said by all the writers who have touched on this phase of the disease to be the malady most frequently complicated with croup, but though I have attended a very large number of cases of measles, I have never seen the disease complicated with croup in any form. Very rarely does croup complicate measles in the commencement; more frequently is it seen at the height of the eruption; and it generally occurs during the stage of desquamation. Spasmodic and catarrhal croup, on the contrary, usually attack during the very onset of measles, and West points out that pneumonia, in all its stages, is far from being unusual, and is a complication especially to be feared in those cases where croup occurs as a secondary affection in the course of measles. Scarlatina is often complicated with a very fatal form of true croup, and as the subject is of importance, I have devoted a brief chapter to the consideration of it, to which the reader is referred. Less frequently than scarlatina, small-pox is complicated with croup, and in a practice among children extending over thirty years, I have seen the complication somewhat frequently in scarlatina and small-pox, but never in measles. In November, 1871, I attended a child, in whom vaccination had been neglected, for confluent small-pox. The case, though very severe, did well till, worn out with watching, the mother did not notice that the child had slipped from the bed to the floor. It lay there for three or four hours, as nearly as could be ascertained, and when taken up pseudo-membranous croup was fully developed, and the child died in twenty-four hours. Again, a young man, unvaccinated and suffering severely from primary syphilis, was attacked with a malignant form of small-pox. Almost immediately the larynx was attacked and the patient died in twenty hours. Steiner has twice noticed pseudo-membranous croup during the height of whooping-cough, and it has also been seen in the course of typhoid fever.

Croup is found in every country and in all climates, yet it is more influenced by peculiarities of country and climate than any other disease of the respiratory organs. Hirsch points out that it diminishes in frequency as we approach the tropics, yet Sir James M'Grigor notes its prevalence at Bombay in the year 1800—but as it attacked adults, it was most likely diphtheritic croup. A cold and moist atmosphere, with rapid alterations of temperature, and the vicinity of the sea, make up the climate in which croup may almost be said to be endemic, and when to these are added an unknown yet very tangible epidemic influence, croup becomes a veritable scourge. According to Bæhr, the flat country extending from Hanover to the North Sea is frequently visited by croup, and he remarks that the winds blowing in this region of country must be possessed of a peculiar nature in order to cause extensive epidemics which sometimes snatch away twenty or more children in a single village. "That croup is caused by a simple cold, is much more easily asserted than proven. The same child has many attacks of violent laryngeal catarrh in the course of the year, but is attacked with croup only during the prevalence of a keen blast from the north." In Scotland the greatest mortality from croup is not found in the extreme north, but on the western and eastern coasts, deeply indented by the sea, which leaves a great expanse uncovered at every tide, and when to these conditions is added the keen easterly winds, croup rages with a great mortality, often exceeding two per cent. of all diseases. The disease is not nearly so frequent in Scotland as it was when Cheyne first wrote, for the low, marshy grounds have been extensively drained, thus affording another illustration of the pernicious influence of moisture. The influence of an equable temperature is strikingly shown by the low mortality from croup in the counties of Wigton and Dumfries in the southeast of Scotland. Here the temperature, though occasionally low, is, on the whole, more equable than in most parts of the kingdom, and the croup mortality is always below one per cent., sometimes touching 0.5. West remarks upon the comparative rarity of croup in towns, and its frequency in rural districts, stating that "out of 100 children dying under five years of age from all causes, more than four times as many will have died from croup in Surrey as in Liverpool, and exactly four times as many as in London." Yet, according to Squire, the highest croup mortality in England is in the populous districts of Lancashire and Cheshire, where, especially in the first-mentioned county, the towns and villages almost touch each other. There can be no doubt that in a dense urban population, with defective drainage and a variable climate, croup must rage with peculiar virulence. "According to the investigations of later years, which indeed are still incomplete, it appears as though the amount of ozone in the air acted an important part as one of the causative influences of croup. This is so much more probable since the amount of ozone contained in the air is liable to the greatest variations during the prevalence of abnormal proportions of electricity such as are apt to be caused by a northwest wind " (Bæhr).

Dr. Elb, of Dresden, questions whether it would not be more correct to ascribe croup, according to the law similia similibus, to the presence of certain component parts in the exhalations from the sea, particularly chlorine, bromine and and iodine, which may act as exciting causes. He proceeds to point out that iodine can produce croupous symptoms, and quotes an observation of Leroy's that, by the accidental respiration of chlorine, symptoms of suffocation were produced, and afterwards concretions were expectorated which very much resembled the false membrane of croup.

The disease may commence suddenly and almost without premonition, but usually it commences with uneasiness and slight shivering, which may not be noticed in an infant. In children of robust constitution, whose general health is good, the disease is apt to come on without premonitory symptoms, but in children of average constitution, the precursory stage is quite distinctly marked. Again, in the debilitated, or in the scrofulous, the grade of inflammation may be low, almost without fever, and exudation closely follows on inflammation. The precursory symptoms are really those of an ordinary catarrh, such as feverishness, sneezing, cough and hoarseness. Bæhr says that " in very rare, or rather exceptional, cases, croup is preceded by a nasal catarrh, which, when present, is a tolerably certain guarantee against the possible occurrence of croup." The *rough cough* and the *hoarseness* are symptoms that should excite attention, for in a young child they are never wholly devoid of danger, and I am satisfied that many lives have been lost from want of attention to this indication. Drs. Evanson and Maunsell urge us to look with suspicion upon these two symptoms—*hoarseness* and *rough cough*—for we can never be too early in our recognition of croup. Note that this cough differs from the cough of spasmodic croup in being *less hoarse and more sonorous.* The eyes are suffused and the child is drowsy, and I am inclined to believe that the latter symptom is much more marked than it is in an ordinary catarrh. The respiration is not irregular except after exertion, and there is slight pain on swallowing, with vague uneasiness in the larynx; but these symptoms are almost wholly unnoticed in infants, and are likely to be overlooked even in older children. The little patient is chilly at times and the chilliness is succeeded by heat of the skin, with lassitude and loss of appetite. The pulse is frequent and a little harder than usual, and the countenance is slightly flushed. This is, of course, a catarrhal fever, but not every catarrhal fever develops into croup, even when the laryngeal complication is quite marked; for, in the words of Dewees, "it would appear that it is not sufficient for the production of croup that the mucous membrane of the windpipe be merely inflamed; but that it requires a modification of inflammation to induce it." The two indications, then, which should suggest croup to the mother are *roughness of the voice* and *hoarse cough.* If at this stage the throat be examined—and an examination should never be neglected—it will be found that, even in cases in which the child has not complained of difficulty in swallowing, there is more or less congestion of the fauces, with exudation of small, pearly, fibrinous spots on the soft palate, uvula, tonsils and posterior wall of the pharynx. These spots are at first mere islands, but they soon spread, and when found in conjunction with the rough and husky voice and the hoarse cough, a morbid state is revealed which should awaken the gravest apprehensions. This entire morbid state, of course, as Dr. Squires points out, follows quickly upon the cause which excited it, and it may last for three or four days, though it very seldom precedes the outbreak of the disease more than twenty-four or thirty-six hours. Even from the very commencement all the symptoms are aggravated at night, and nocturnal exacerbations are the rule throughout the disease. As Dr. Charles West accurately remarks, " thirty-six hours seldom pass without the supervention of some symptom which, to the well-schooled observer, would betray the nature of the coming danger." But the precursory stage may be absent, and in the robust or in the scrofulous the laryngeal inflammation followed, or rather accompanied, by exudation, may be the first intimation of danger. Professor Wood says: "I once attended the case of a little girl who, when first visited, was running about the apartment with no other apparent disease than a whispering voice, and perhaps some little difficulty of respiration; yet she was at that moment almost as surely condemned to death as though she had been already in the last stage of the disease; for the membrane was already formed, and no efforts could prevent its fatal progress." The writer has attended a number of cases in which, after some over-exertion at play, or after exposure to cold, children were attacked without warning; but this sudden onset is of rare occurrence. Sometimes pseudo-membranous croup is developed in the course of spasmodic croup—especially if the child has had repeated attacks—and the possibility of this should be kept in view by the mother and by the physician.

The outbreak of the fully developed disease almost always takes place about midnight. The early part of the night may be passed in quiet sleep, but the child is suddenly aroused by a severe paroxysm of cough, or, more rarely, by a series of coughs, gradually increasing in number and violence. This second stage is marked by a change in the character of the cough, which has a ringing, brassy clangor, which can never be forgotten when once heard. Evanson and Maunsell say that the cough is sharp and ringing, as if passed through a brazen trumpet, and Bæhr compares it to the bark of a watch-dog, but all comparisons poorly picture its ominous, ringing resonance. This change in the character of the cough heralds a change in the respiration, which becomes prolonged and stridulous—a loud rattling noise succeeding each inspiration as well as each paroxysm of cough. In the most severe cases this loud, rattling noise accompanies the expiration as well as the inspiration. The inspirations are audible, wheezing and much longer than normal, and the respiratory acts are greatly more frequent than in health, from 28 to 36 to the minute, occasionally as high as 48. The paroxysms of dyspnœa are of the most frightful character. The child sits up in bed, stretches his head backwards, and instinctively does all he can to force air through the narrowed glottis. The hoarseness, which was present during the first stage, is now replaced by an almost complete suppression of voice, which falls to an almost inaudible whisper. The cough loses its ringing, sonorous sound and becomes dry, husky, and apparently confined to the throat. It is distinctly paroxysmal, and though it is sometimes frequent, in other cases it occurs at long intervals, and I have noted that the frequent cough is a more favorable sign than the rare cough, while the complete or almost complete suppression of cough is a very bad sign indeed. As the second stage progresses, the cough becomes shorter and more smothered, till as Dr. Meigs remarked, "it might very well be called whispering." The breathing now becomes still more difficult, the cough assumes the muffled and husky character, the gestures of the child indicate pain in the throat or upper part of the sternum, the face becomes swollen and darkened, the anxiety and unrest becomes excessive, and all the symptoms indicate approaching suffocation. The little one starts up in bed and begs piteously to be taken in his mother's arms, immediately he entreats to be put back to bed again; he grasps his windpipe as if he would tear out the obstruction to respiration; he tosses about in his crib, catching at its sides in his agony; the face is livid and distorted; the red and swollen eyes almost start from their sockets; the veins in the head and neck are thick and blue and cord-like; cold perspiration covers the brow, yet the cough, in spite of the most desperate exertions, is still soundless, accompanied by the expectoration of a very little tenacious mucus, mingled with froth. "In a word," says Steiner, "we have before us the heartrending picture of a child nearly suffocated, tortured with the death-pang; a picture which draws out all our compassion, and brings home to us, as few other diseases do, the painful side of our calling."

In the early part of the attack there is no expectoration, or perhaps a little viscid mucus; but during the second stage there may be expectoration of false membrane in small pieces, mixed with ordinary mucus. Dr. Meigs says that to "detect the membrane, the substance expectorated or vomited ought to be placed in water, when the former detaches itself from the mucus and other matters and is easily recognized." When first thrown out on the mucous membrane of the pharynx and larynx the yellowish exudation is quite fluid, but it soon coagulates, and when discharged it is in shreds of various sizes and thickness, or complete casts of the larynx may be ejected with immediate relief of the symptoms. Quite often I have seen membranes of the consistence of the upper layer of thick cream, and I have noted membranes as dense and firm as kid leather; sometimes the creamy membrane comes away mingled with shreds of the denser type. Valleix detected the membrane in 26 cases of 51, and Drs. Meigs and Pepper write, "Of the 35 cases observed by ourselves, it was expelled by vomiting or coughing in 12; in 21 none was ejected, though its presence in each case was proved by the character of the symptoms and by its existence in the fauces, by autopsy or by the operation of tracheotomy; in one there was expectoration of masses of viscid, yellowish fibrin, though none of membrane; and in tone there was no positive evidence of its existence." At times a fragment of false membrane is detached, wholly or in part, from the laryngeal mucous membrane, and is carried below the vocal cords, causing a long-drawn, suffocative paroxysm, which may prove fatal unless, by a desperate effort, the membrane is dislodged. When the loosened membrane is of small size it makes a flapping noise, easily recognized by the stethoscope.

It will be noticed that all the symptoms remit, but are

never wholly absent, and the slightest cause, as taking a little food or saying a few words, causes an immediate return, with increased violence. At first the fever is slight, and in many cases it is altogether absent, but in the second stage fever is almost invariably present, and in general terms it may be said to be high in proportion to the extent and intensity of the local inflammation. The pulse, which was full and hard during the first stage, and from 110 to 125 to the minute, in the second stage is slightly more frequent, rising 20 to 30 beats during the suffocative paroxysms and falling as much during the remissions. If the disease should extend to the bronchial tubes, at once the pulse increases in frequency. While the disease is, as a rule, marked by remissions in some cases, in the words of Professor Wood, it "marches directly onward to suffocation almost without paroxysms." The suffocative attacks seem an age to the anxious medical attendant and still more anxious mother; in reality they last but three to six minutes, rarely a quarter of an hour, and they commonly end in a certain relief, marked by a brief slumber, but the wheezing inspirations tell of the continued presence of a terrible danger. As morning approaches there is a longer remission of all the symptoms, even of the loud rattle, and a sleep of some length is obtained ; but I am not prepared to say with Bæhr, "in the morning the little patient may feel quite well, except perhaps a little weak and languid." That is not my experience, for next day I always have very sick patients on hand, in whom a mere remission of the disease is present, thus affording precious time for further treatment. I grant that croup, as a general rule, shows a decided remission in the morning, which sometimes almost amounts to an intermission ; certainly the respiration is more free and the voice returns, and the fever, too, abates, and even the cough is less frequent. But the cough has a reedy, piping tone which suggests trouble during the coming night, the fever shows that the local inflammation still exists, and on examining the pharynx it will be found that in a majority of cases pearly islands of false membrane tell of still greater deposits in the larynx. Note that this remission is the time for successful treatment. Neither do I agree with my distinguished German colleague when he writes: " Up to this period (the morning after the night in which the fully developed disease appeared), croup resembles an ordinary attack of laryngitis so perfectly that it is often impossible to distinguish one from the other. This uncertainty and vagueness of the symptoms may continue during the second and even third night, although the croupy character of the attack becomes more and more marked as the disease progresses on its course." So far as my experience goes, the characteristics of this truly frightful disease are present from the time of its first outbreak. Only in comparatively rare cases can there be any doubt as to the diagnosis on the morning after the attack. One little-noted characteristic of the disease is apt to throw the practitioner off his guard, that is, that the first paroxysm is often followed by a remission so nearly perfect that the most careful auscultation is needed to prove the existence of the disease.

Let us pause here and consider the *mechanism of the disease*, for much depends on a knowledge of it. The dyspnœa of croup has, from the first, fixed the attention of medical observers, and the old view, once universally held, is that it is mechanically produced by the croupous membranes. Later, the idea of spasm of the glottis gained a number of adherents, and Billard and other French writers still maintain this hypothesis. Another explanation given by Bretonneau, and still held by a small number of practitioners, is that the dyspnœa is caused by the difficulty with which the secretions of the bronchi are forced through the glottis narrowed by the deposit of false membranes. Rokitansky maintains that "the infiltrated, pale, relaxed muscular tissue, in croupous inflammation, is stricken with palsy," and he looks upon dyspnœa as a result of the paralysis of the laryngeal muscles. These views of Rokitansky are supported by Schlautmann and von Niemeyer, and the latter points out that section of the par-vagum nerve in young animals furnishes absolute proof that paralysis of the muscles of the larynx produces dyspnœa; nay, the dyspnœa arising in consequence of this experiment bears so strong a resemblance to croupous dyspnœa, is attended by such similar long-drawn, whistling inspiratory efforts, and other signs, that the similarity of the two conditions must strike the most indifferent beholder." Dr. von Niemeyer further remarks that in paralysis of the laryngeal muscles the inspiration is laborious and prolonged, while the expiration is free and almost normal, and he sums up his views on this interesting topic by stating that paralysis of the laryngeal muscles causes laborious and whistling inspirations with free expiration, while the narrowing of the glottis by croupous membrane is really an interference with both entrance and exit of air, hence the difficulty in both inspiration and expiration. There is truth in all these apparently discordant views, but their supporters have been too one-sided, for any intelligent physician who carefully studies a few cases of this disease will see that the dyspnœa has three distinct sources —*mechanical, spasmodic and paralytic*—and of these the most influential is certainly the mechanical one—the narrowing of the larynx by the swelling of its walls and the false membrane deposited on its surfaces. Paralysis of the laryngeal muscles is present in many cases, and certainly spasm of the glottis is common, and these three factors make up the dyspnœa of pseudo-membranous croup.

The vocal cords swell and thicken, and they are coated with a very delicate false membrane, which gradually increases in thickness, and, as a result of these morbid changes, the vibratility of the cords is altered. Hence the characteristic hoarseness. The older practitioners considered that the cough was caused by spasm of the glottis, but it is now known that the tone of the cough, like the changes in the voice, depends on the same swelling and thickening of the vocal cords, resulting from the deposit upon them of the characteristic false membrane. As a result, the cords are greatly less mobile than in health, and hence the voice is, from the first, rough and harsh, then crowing and barking, and finally, it is suppressed. Quite likely, a partial paralysis of the muscles which open the glottis—the crico-arytænoidei postici—is an essential ingredient in the changes in the voice and in the tone of the cough.

As a rule, then, the patient on the morning after the first nocturnal attack has some degree of hoarseness of the voice, with a hoarse and resonant cough, and I have noted that a free and frequent cough gives better promise of recovery than an infrequent or suppressed one. The pulse will be fuller and more frequent than natural, and the temperature will be a degree and a half, or so, higher than the normal. As night approaches the breathing becomes loud, difficult and wheezing, the cough is of the same character as on the previous night, only less sonorous and more distressing, and already the patient feels the sensation of impending suffocation. When the paroxysm of cough approaches the little patient rises in his bed, clutches at the mother with a dreadful energy, or falls down as if convulsed. But little expectoration attends the cough, at best a small amount of glairy mucus, sometimes blood-streaked, is discharged. The pulse is hard and frequent, though the temperature is somewhat lower than on the previous night. The face is flushed and swollen, and the voice, hoarse at the beginning of the night, is often almost inaudible by morning. At the very latest the disease is at its height by the close of the third day, but often long before that time the intensity of the attack has developed the third stage—the stage of asphyxia, or rather of threatening suffocation. As this stage approaches, the remissions between the paroxysms grow shorter and shorter, till the paroxysm is continuous or nearly so. The voice is whispering or entirely suppressed; the cough is dry, stifled, infrequent or entirely absent; the respiration is slower and more convulsive from the mechanical obstacle to the entrance of air; and both inspiration and expiration are marked by a loud stertorous noise. The child is drowsy, and from that ominous slumber it starts in terror and grasps at the poor throat. The cool skin is now covered by clammy sweat; the pulse is too rapid and too weak to be counted; the respiration is so superficial that the dyspnœa is scarcely noticeable, and the loud stridor is no longer present; at each inspiration the larynx is drawn downwards towards the sternum; at times a desperate rally is made and the child struggles hard for breath, but the face becomes cyanotic, the extremities cold, and death takes place amid coma or convulsions, which sometimes strikingly resemble those of spasm of the glottis. The closing symptoms are caused by the overloading of the blood with carbonic acid, as much as 3.27 per cent. has been found in the expired air.

When a favorable change takes place it is usually before the appearance of the third stage, for a very small number of these recover. Usually a discharge of false membrane is one of the first signs of amendment, and this discharge is sometimes preceded by a sound in the larynx as of loosened membrane flapping to and fro with the respiratory movements. Generally the expectorated matter is simply tough, and whitish shreds mingled with muco-pus—only rarely is a cylindrical cast of the affected parts ejected. Bæhr remarks that when a cylindrical cast is thrown off, it is not safe to regard the danger as entirely over until at least two days have elapsed without any trace of a renewed exudation having been perceived. The cough becomes milder, the breathing easier, the larynx clearer, the fever ceases, the skin becomes moist and soft, and slowly but surely the child enters upon convalescence. The amendment may be sudden or gradual, and the writer has observed that a sudden amendment is more likely to be followed by a relapse than a gradual one. Dr. West thus describes a striking phase of the malady: " The mitigation of the disease may be accompanied by great drowsiness, which, however, does not excite alarm, since it is very naturally attributed to the exhaustion produced partly by the disease, partly by the remedies. During sleep the respiration is deep and tranquil, like that of a person in a sound slumber; it is, indeed, attended by a kind of wheeze, but presents little of the croupy stridor; and when awake the child is quite sensible, and even cheerful. After a time, however, it becomes difficult thoroughly to rouse him; his pulse grows more rapid; the moisture on his skin changes almost imperceptibly to a cold, clammy sweat, and convulsive twitchings of the angles of the mouth occasionally disturb the repose of the features. Silently, but surely, the exudation has been making progress, and when the alarm is taken it is too late; the stupor deepens and the child dies comatose, or rouses, only to spend its last hours in the vain struggle for breath, and embittered by all the painful circumstances which ordinarily attend the suffocative stage of croup."

At times, the course of this disease is extremely rapid. Much depends on the vitality of the child and a good deal upon its docility. In cases where the remissions are completely absent, the disease marches on to a fatal termination in from twenty to thirty-six hours. I attended one fine boy, seen after an illness of only eight hours, who died at the close of the eighteenth hour; Vogel says that the shortest time he has known, from the invasion of the malady till death, was twenty-four hours, and Dewees has seen it run its course in a few hours. Generally, the disease lasts from three to five days, and Dr. Copland has noted that a fatal issue is most common on the fourth day. Dr. Craigie says that it is never protracted beyond the eleventh day, but Steiner has seen in a child, five years of age, false membranes upon the bronchial mucous membrane even forty-nine days after tracheotomy, and he is certain that exceptionally the disease may run three, four or more weeks. He adds, "the longest duration is in the ascending croup." Drs. Meigs and Pepper's cases lasted from three to fourteen days, and Vogel's longest case died after eight days illness.

No very large number of thermometric observations have been made, and they all confirm the remark of Wunderlich, that in no other diseases has the temperature so little significance as it has in croupous and diphtheritic affections.

During the incipient stage of the disease, the thermometer shows a temperature varying from 99° to 100°; on the night of the outbreak it may be 102°, rising to 103° or so during the paroxysms of dyspnœa, and falling with their departure. During the next day the temperature is 100° to 101°—the higher the temperature the more severe the attack during the approaching night. But during that second night the temperature rarely equals that of the first, and it continues to decline unless bronchial or pulmonary complications appear, when it may rise to 104°, 105°, or even 106° in exceptional cases. I have attended a number of cases in which the temperature never rose above 101°, and then I had reason to remember the warning of Wunderlich, "*moderate or even normal temperatures do not give the slightest guarantee for a favorable termination.*" Dr. Squire remarks that "a high temperature at the very outset may point to one of the exanthemata, its persistence to diphtheria."

The larynx is the seat of pseudo-membranous croup, but the larynx is very rarely the only organ affected, the inflammatory irritation usually passing down the trachea and bronchial tubes, even extending to the bronchioles. The older writers divide croup into "descending" and "ascending," but for many years there has been a strong disposition to question the existence of the ascending croup, and many physicians in large practice among children have never seen a case. Steiner, however, has attended four well-marked cases of ascending croup. " In each case the disease began with slight febrile symptoms, more or less cough of a painful character, and dyspnœa. After from four to six days, while the voice was still completely sonorous and without any indication whatever of laryngeal obstruction, croupous membranes were expectorated. Towards the end of the first week, and in two of the cases on the fourteenth day—the fever still continuing—hoarseness occurred, followed by laryngeal stenosis in its full intensity, and, shortly before death, by the deposit of false membrane upon the faucial mucous membrane. In each case the disease was ascribed to a severe chill; three died, and one, a girl five years of age, recovered." He adds, " rare as such cases certainly are, their occurrence is unquestionable."

Trousseau says that "the admirable diagnostic methods — auscultation and percussion — given by Laennec to the profession for the general good, and of which no one is allowed to be ignorant, are in our hands what the telescope and magnifying-glass are in the hands of the astronomers and the naturalist, instruments intermediary between external objects and the mind," and it is precisely in the laryngeal diseases of children that the stethoscope is too much neglected. I have found Cammann's stethoscope, especially the single one, immeasurably superior to the old instrument, stigmatized by Abernethy as being "a piece of wood with a patient at one end and a fool at the other." Especially in croup has it served a good turn, though it is not an infallible means of diagnosis, for Dr. West says that he noticed on one occasion those changes in the tracheal sound which are supposed to indicate the presence of a very extensive deposit of false membrane, although no false membrane was either expectorated during the patient's lifetime, or discovered in the inflamed larynx or trachea after death. He adds " we must conclude, therefore, that the changes in the tracheal sound do not afford absolutely certain evidence of the existence of false membrane, and that still less can they be regarded as safe criterions of its extent." At the commencement of the disease, laryngeal auscultation simply reveals the characteristic stridor, though the air enters easily, and when false membrane forms, the sound in the larynx and trachea usually becomes less stridulous and more sibilant, though, as already remarked, there are exceptions to this rule. Barth and Rogers state that when false membrane exists, *tremblotement* —a trembling, vibratory murmur—is present, and that the extension of this sound downward demonstrates the extension of the disease. Unfortunately, the *tremblotement* caused by the presence of mucus in catarrhal croup cannot be distinguished from the same sound caused by the false membrane of the more malignant disease. In all cases the information derived from auscultation must be compared with the vital symptoms present, and the course of the disease must be carefully investigated.

During the first stage the inspiratory sound is prolonged into a harsh and prolonged stridor; the expiratory sound is also prolonged and harsh, but low in pitch, while the weakened respiratory murmur is effectually masked by the shrill, laryngeal stridor. Still, a certain well-defined mucous rhoncus is heard over the larger bronchial tubes, especially at the moment when the child makes a very deep inspiration. No dullness on percussion is present, and the entire respiratory movements are notably deficient, even at this early stage, and the walls of the chest are never fully expanded. During the second stage the sibilant inspiration is distinctly heard during sleep, and especially on waking, but it has lost the loudness and the persistence which marked it during the first stage. Lastly, during the final stage of rapidly advancing apnœa the characteristic tracheal sound is audible all over the trachea during expiration. Dr. Hartshorne remarks that a mucous râle, sufficiently tremulous to be audible without the stethoscope, is usually a very favorable sign, as it almost invariably indicates that the mucous follicles are throwing out their secretion between the mucous membrane and the false membrane—a process on which the cure at this late stage greatly depends.

The laryngoscope can hardly be used with croup patients. Even Steiner, with all his skill, is forced to acknowledge that it is "almost impossible." Münch gives us the following description of his laryngoscopic examination of a boy of ten years suffering from croup: "The mucous membrane of the larynx was much reddened; a marked membranous deposit covered the aryteno-epiglottidean ligaments, and still more copiously the vocal cords; the glottis was narrowed, partly by the deposit on the vocal cords and partly by the paresis of the dilator muscles—the posterior crico-arytenoid.

Later the whole larynx appeared to be covered with membrane; at the same time it was noticed that the edges of the vocal cords were apparently agglutinated to each other at various points by a layer of fluid exudation. Subsequently the deposit disappeared under the continued use of caustics, but was renewed daily, until finally only a thin, gauzy membrane was noticed, which returned again and again with great obstinacy, especially upon the vocal cords. The vocal cords ultimately resumed their function, and manifested considerable vitality, even while some of the membrane remained. By the sixteenth or eighteenth day the normal white color of the cords was restored, and here and there a reddish streak was all that could be noticed."

Wherein lies the difference between the ordinary inflammation of mucous membrane and the exudative form of inflammation? Why is it that one child has, after exposure to cold, a simple catarrhal croup, while another, after the same exposure, has pseudo-membranous croup? Dr. Searle of Brooklyn, remarks, "wherever the distinction may lie pathologically, the fact is certain," and I shall endeavor, as correctly as may be, to define the condition which lies behind, not only the symptoms, but behind the proximate cause which gives rise to the symptoms.

There is, then, an increased proportion of fibrin, or of fibro-albuminous matter in the blood, and this is considered by Dr. Cheyne to be analogous to the exudation of the inflamed pleura or peritoneum. This fibrinous material has a kind of inherent tendency to organization, and this imperfect textural development appears to set in with the process of coagulation. "Examined with the microscope, they present a laminated basement, and one splitting into fibres, flattened or roundish, rough and firm, or resembling organic muscular fibres; or else, a membranous basement invested with delicate, wavy fibres, upon which, among elementary granules, are seen numerous round, black-edged nuclei, sometimes rod-shaped, or drawn out into fibres, and again, more especially in the moisture poured out, dull, round or oval nuclei, and analogous cells." (Rokitansky.) This first variety of pseudo-membrane is a tough, elastic, polished membrane, quite similar to serous membrane in appearance, and very like moist kid leather. At other times the fibrinoid is of a dullish white color inclining to yellow, and including blood-serum and blood-corpuscles, sufficient to give a reddish hue in places. " Microscopically examined, the coagulum presents a stratiform or fibro-laminated basement, or else a faintly striated membrane, both being, however, opaque, owing to delicate granulation. Upon this, as also in the serum, are seen a vast number of nucleus-like formations of developed, dull, granulated nuclei, and of similar more or less developed cells. Frequently the coagulum appears to consist altogether of the two last-mentioned elements, with a proportion of granulated structure." (Rokitansky.) Again, the fibrinoid may be pus-like, of a greenish-yellow hue, with little tendency to organization, and but little adhesive power. These three varieties rarely occur singly and alone, but they are intermingled in varying proportions. The first mentioned is the most dangerous, as it approaches the nearest to organization ; and the last mentioned is analogous to the matter of pyæmia, for it includes pus-nuclei and pus-cells in its meshes. The difference between these varities of fibrinoid can be readily detected with the naked eye.

When the mucous membrane of the larynx becomes the seat of inflammation or of congestion—for there can be little doubt that in croup the primary morbid change is often congestion—this fibrinoid or fibro-albuminous matter exudes from the distended capillaries, and the change of temperature and the passage of air over it aiding its inherent tendency to imperfect organization, it is soon formed into a false membrane. As the disease advances, the mucous' follicles secrete a copious muco-purulent. fluid which is poured out between the mucous membrane and the false membrane, loosening the latter, so that there is a certain tendency toward recovery even in the most severe forms of true croup. This congestive or inflammatory action, with its accompanying exudation, may go on undetected for some little time till the engorgement becomes so great as to interfere with the passage of air through the glottis, or till the more or less violent laryngeal spasm directs attention to it. The writer is strongly of the opinion that while simple spasmodic croup, almost destitute of inflammatory action, stands at one end of the scale of morbid action, at the other extremity is pseudo-membranous croup, which may be almost wholly destitute of laryngeal spasm: that though well-marked typical cases exist, which can be readily diagnosed, yet in the middle of the scale we find it extremely difficult to decide as to the presence or absence of false membrane; and, lastly, that a case which apparently commences as spasmodic croup, may, under certain conditions, take on inflammatory action with its attendant exudation. The practical lesson is to prescribe for even mild cases with care, and constantly to keep in view the possibility of the occurrence of the much dreaded pseudo-membrane.

Dr. Squire asserts that "intense redness of the mucous membrane is persistent after death," but even when the redness, as observed by the laryngoscope, has been intensely bright, at the post-mortem examination the hyperæmia may have entirely disappeared or be scarcely noticeable. Swelling is rarely found, as it, too, disappears with the extinction of life, though sometimes the upper orifice of the larynx is diminished by the swollen aryteno-epiglottidean folds. Dr. Craigie asserts that croupous inflammation is but seldom observed to affect the laryngeal mucous membrane, and says that when it does so, it is to be viewed as a complication not essential to genuine croup; while, on the other hand, Guersant says that the characteristic membrane is never entirely absent from the larynx. Here Dr. Craigie is unquestionably in error, for the larynx is always affected in croup, though, as Rindfleisch remarks, a croupous inflammation, confined throughout its entire course to the larynx, is of rare occurrence.

In general terms it may be said that in two-thirds of all the cases the disease is limited to the larynx and trachea, while in the remaining third the inflammatory irritation extends to the bronchi; though it does not follow that false membrane is formed there. "The implication of the trachea and bronchi is, at least with us, very common; in fifty-five autopsies of children I found that in thirty-one the croup had extended to the larynx, trachea and bronchi, with casts even in the smaller tubes; in nineteen the false membranes were limited to the larynx and trachea with purulent or muco-purulent secretion on the mucous membrane of the bronchi, especially those of the first and second order; in the other five cases croupous deposits were present only in the throat and larynx, with muco-pus in the trachea and bronchi. It is to be particularly noticed that in *all these cases* false membrane was demonstrated in the laryngeal cavity, and it is safe to say that the absence of exudation, to which some are so ready to appeal, is unquestionably the very rare exception" (Steiner). Again, it may be confined to the glottis and it may line the entire larynx, dipping into the ventricles so as to form an entire cast of the organ; in very severe cases it extends to the minutest ramifications of the bronchial tubes, and this seems to be more common on this continent than in Great Britain. Professor Wood has seen a case in which the false membrane lined the upper portion of the bronchial tubes, the entire larynx and trachea and the pharynx as low as the upper part of the œsophagus. But the favorite situation of the false membrane is on the vocal cords, and, as a general rule, the coating is thick in proportion to the duration and severity of the attack. Dr. Cheyne compares the tubes of false membranes from the bronchial tubes to macaroni boiled in milk, and, in curious anticipation of Dr. Craigie, he says that in none of the cases seen by him was membranous exudation observed on the laryngeal mucous membrane, adding, that if the inflammation extended to this part, it was only slight, and its effects were seen in a little puriform fluid on the membrane of the cricoid or thyroid cartilages.

Sometimes the false membrane adheres closely to the mucous membrane, but it is generally more or less loosened from the action of the muco-purulent fluid already mentioned, and this loosening is, in the words of Rindfleisch, "*a property on which all our therapeutic measures, inadequate as they are, repose.*" After the first membrane is thrown off, a second succeeds it, and then a third, till death takes place or recovery ensues from the false membrane ceasing to form, and when this takes place it is found that the mucous membrane is but little injured, in fact it is often quite normal. There are great differences in the thickness and consistence of the false membrane ; it is sometimes of a gauze-like tenuity, while at other times it is one or two lines in thickness, the usual thickness being about half a line ; as already mentioned it may be like the viscid layer which forms on the surface of a bowl of cream, and it may be a tough, compact, leather-like fibrin resembling a fragment of wet kid glove. Almost invariably, the edges of the membrane are thinner and softer than the more central portions, and the side in contact with the mucous membrane is softer than the side exposed to the air. When it extends to the bronchial tubes, Rokitansky remarks that the tubular exudations from the larger bronchi present a calibre inversely proportional to their thickness, and those thrown off from the finer ramifications occur in solid cylinders. Professor Wood remarks that in the larynx it is said to be less firm than in the trachea ; while Professor Gross asserts that it is generally much stronger, more tenacious, and more firmly adherent in the larynx than in the trachea and bronchial tubes. " The characteristic feature in the morbid anatomy of laryngeal croup is due to the fact that the mucous lining of the larynx agrees in its structure, partly with that of the pharynx, partly with that of the trachea. Both surfaces of the epiglottis, and the true vocal cords, are coated with a laminated pavement-epithelium, which is not marked off from the connective tissue by any homogeneous basement-membrane. Hence, the false membranes adhere more firmly to these than to any other points in the interior of the larynx. How often do we find, in making a post-mortem examination, that the tracheal false membrane, continuous with that of the laryngeal funnel, is quite loose as far up as the rima glottidis, where it is firmly attached ; and we feel sure that its spontaneous detachment at this point would have required a very long time for its accomplishment." (Rindfleisch.) The color of the denser membrane is of a pearly, grayish white, while the more diffluent membrane is of a yellowish white.

Small quantities of carbonate of soda and phosphate of lime have been detected in the false membrane, and it is soluble in acetic acid and alkaline solutions, especially in lime water, in short, in all its chemical relations it closely resembles coagulated fibrin. Examined microscopically, according to Steiner, it is found to be composed of amorphous or fibrillated fibrin, in which numerous young cells are entangled. Squire says that it is not simply fibrin, but that " it consists of effused lymph, in which the presence of albumen can always be chemically demonstrated ; microscopically it is a mass of cystoid corpuscles."

Is the false membrane susceptible of organization? Generally speaking, it gives no indication of such an attribute, and yet Rokitansky thinks that an effort at organization occasionally takes place. " The surface next to the mucous membrane is frequently marked with red streaks and dots, consisting in part of blood adhering to the surface, and in part, as found on closer examination, of straight or tortuous vessels, or of small, roundish, extravasations, from which currents of blood are seen to emerge in an arborescent and radiating form." Professor Hasse remarks that the effort at assimilation is, in some instances, very perceptible in the appearance of stellated ecchymoses and bloody streaks on the surface of the false membrane, facing the mucous membrane.

The mucous membrane subjacent to the false membrane seldom presents the appearance of severe inflammation, though it may be red, purple, or even blackish in color, and these tints are in spots or patches, which are sometimes arranged in irregular stripes. The mucous membrane is sometimes, but rarely, in a state of gelatinous softening, and thickening is still more rare. West has observed ulceration in one acute case, but frequently in cases of secondary croup, probably diphtheria. But, on the whole, the mucous membrane producing the croupous membrane remains, as Squire remarks, "singularly free from pathological injury." At an advanced period of the disease the redness may disappear, when the mucous membrane regains its usual pale color. The trachea and bronchial tubes are usually reddened, even though the disease has not extended to them, and the bronchial tubes contain a yellowish, puriform fluid, which has doubtless passed downward from the seat of morbid action. A certain degree of pulmonary congestion is an almost inevitable result of croup, and the same may be said of the vesicular emphysema, which results from the extraordinary efforts to breathe, which sometimes brings on laceration of the pulmonary vesicles. The same desperate respiratory efforts often cause congestion of the brain and even effusion of serum into the ventricles.

To those physicians who believe that there is but one kind of croup, clearly but little diagnosis is necessary. Thus the celebrated Dr. Dewees speaks of "a distinct species of croup, namely: the spasmodic, a kind we have never witnessed," though Drs. Meigs and Pepper, who practice in the same city that was honored by the residence of Dr. Dewees, declare that they meet with six cases of spasmodic croup for one of pseudo-membranous. Again, Dr. Robert C. R. Jordan, Professor of Diseases of Children, Queen's College, Birmingham, England, writes as follows: "In all my own early teaching it was strongly impressed upon me that "croup" was always a membraneous exudation in the larynx or trachea, that it became to my mind a great difficulty to throw off the trammels of this old belief, and it was long before I could feel fully persuaded of what I now know to be the truth—namely, that the majority of the cases usually called by this name have no false membrane formed at all, but that their essential nature is an inflammation of the mucous membrane of the larynx and trachea, accompanied with secretion of tenacious mucus, and also considerable swelling caused by effusion into their submucous areolar tissue. They are, in fact, catarrhal inflammations of the larynx and trachea. All other cases where exudation is really present are diphtheria ; and it is in this sense and with this definition only that we can regard croup and diphtheria as two distinct diseases."

The diagnosis between pseudo-membranous croup and catarrhal croup will be found in Chapter V, that between croup and spasm of the glottis in Chapter IV. It remains, then, to give the diagnosis between spasmodic croup and pseudo-membranous croup, for a diagnosis can certainly be made in spite of the confident assertion of. Dr. Maunsell that " there are no means of distinguishing between the two affections (if two distinct affections exist), beyond the degree of violence of the symptoms."

The attack of spasmodic croup, then is, as a general rule, sudden and startling, while the invasion of pseudo-membranous croup is insidious and creeping. In spasmodic croup the voice is hoarse but never whispering, save during the height of the attack, while in pseudo-membranous croup, the voice, at first hoarse, soon becomes whispering, and finally is entirely lost. The cough of spasmodic croup is rough and hoarse throughout, while in pseudo-membranous croup the cough is rough and hoarse at first, infrequent further on, and finally is quite suppressed. In spasmodic croup the suffocative attacks generally occur at the beginning of the disease, in pseudo-membranous croup they come on at an advanced stage. In spasmodic croup great dyspnœa is very rare and it never persists, but in pseudo-membranous croup dyspnœa is an essential feature of the disease. The respiration of spasmodic croup is stridulous and difficult only during the paroxysm, but almost natural in the interval, while in pseudo-membranous croup the respiration, at first almost normal, becomes very nearly permanently stridulous. In spasmodic croup the fauces are quite clean or a slight redness may be present, but in pseudo-membranous croup a fibrinous exudation is quite common. In spasmodic croup the fever is very slight and may be altogether absent, and it may only appear during the paroxysm, while in pseudo-membranous croup the fever is quite high. After the paroxysm of spasmodic croup the child is quite well and the fever departs, but after the first paroxysm of pseudo-membranous croup the child is quite ill, with high fever, stridulous breathing and hoarse cough. If the paroxysm of spasmodic croup returns the second night it is less severe than it was the previous night, but the second nocturnal paroxysm of pseudo-membranous croup has increased dyspnœa and threatening suffocation. Spasmodic croup rarely lasts more than three days, and good treatment reduces this to thirty-six or forty-eight hours, and it is not followed by hoarseness, while pseudo-membranous croup is rarely of shorter duration than five or six days, and hoarseness is apt to continue for two or three weeks. A child may have repeated attacks of spasmodic croup, but, as a general rule, true croup does not recur. Lastly, spasmodic croup is hardly ever fatal, while pseudo-membranous croup, in spite of the best treatment, is frequently fatal.

Pseudo-membranous croup is always a serious disease, but the homœopathic physician need not assent to Vogel's maxim, "*the prognosis in well-declared croup may be set down as fatal,*" or even to Sir Thomas Watson's well-known dictum, "*the prognosis can never be better than doubtful,*" though that, after all, is merely the legitimate result of a treatment comprehending *blood-letting, tartarized antimony and calomel*, said by the last-mentioned eminent authority to be "the remedies that most require consideration." Almost all physicians—especially those of European education and experience—will assure you that, under any circumstances, a majority must die, but with a thorough knowledge of the pathology of the disease and of the admirable therapeutic agents which homœopathy places at our disposal, the writer believes that a majority will live.

Guersant says that it is "generally fatal," adding that it is scarcely possible to save two in ten, while Rilliet and Barthez state that "its common termination is in death." Steiner thinks that the prognosis is "almost always dismal, a fatal result being almost the rule, for in tracheotomy alone there seems any chance of recovery." Felix von Niemeyer considers it "one of the most formidable of diseases," and Squire thinks that "the slightest cases of croup furnish grave cause for anxiety." Maunsell is almost the only writer who takes a really cheerful view of the matter, and he affirms that "it is remarkably within the control of art"—very true of spasmodic croup, but not quite so true of pseudomembranous croup. Dr. Squire speaks of "the hopeful conjecture of Dr. Wood, of Philadelphia, that one case in fifty only is fatal," but Dr. Squire overlooks the fact that the great Philadelphian is speaking of the "ordinary croup of this country," which is catarrhal and spasmodic in its nature. Meigs and Pepper lost sixteen out of thirty-five cases, though they have attended two hundred cases of spasmodic croup without a death.

The danger is great in proportion to the youth of the patient. A child a year old has less chance than one of five years, and, as already stated, the disease is more fatal in boys than in girls. Vogel's experience is that children who have passed their seventh year may survive attacks of croupous laryngitis of the utmost intensity.

Very much depends upon the stage of the disease at which the patient comes under treatment, and quite as much depends upon the physician possessing an accurate knowledge of the disease before him. If no efficient treatment is adopted till the disease is fully developed and the false membrane formed, the prospect of cure is much diminished; but if, on the other hand, it is recognized from the commencement, and skilful medical attendance is joined to careful nursing, the chances of recovery are better. Pneumonia aggravates the danger, and when the bronchi become implicated in the disease, the prognosis is very grave, though it is well to remember that in bronchial croup the membrane is less firmly adherent than in the laryngeal and tracheal forms. Although the general symptoms should be duly weighed, especial attention should be paid to the local symptoms, and to the frequency of the paroxysms. It is unfavorable if the stridulous sound is heard both in inspiration and expiration, and complete extinction of the voice and suppression of the cough are most ominous signs. Sir Thomas Watson remarks, "we begin to despair when the lips are turning blue, the skin is losing its heat, the pulse is already feeble and intermitting, and the little patient is drowsy or comatose." On the other hand, the favorable signs are diminution of the stridulous respiration ; return of the voice, even though it be hoarse ; looseness of the cough with expectoration of muco-purulent matter mingled with fragments of false membrane ; and decrease of the dyspnœa. Dr. Meadows observes that "sometimes just as the active signs of the attack are subsiding, a relapse takes place, and the condition becomes a much more alarming one; this tendency to relapse should make our prognosis guarded for at least two or three weeks, and particularly in weak, delicate and irritable children." In like manner, Dr. Charles West says that much caution must be exercised in drawing a favorable conclusion from a diminution of the severity of the symptoms, until such improvement has continued for twenty-four hours at least ; and I can most cordially endorse Dr. J. F. Meigs' axiom, "the case should not be abandoned as hopeless until life is actually extinct." Very cheering, too, are the words of Maunsell, "children have recovered from the most hopeless condition, and we should never despair of a sick child."

Children who show any tendency to croup should not go out of doors when cold east winds are blowing, and when such children go out, even in moderately cold weather, they should be warmly and comfortably clad, especially about the feet and neck. In addition to this precaution the neck and chest should be systematically sponged with cold water every morning, and in addition, gargles of cold water should be used two or three times a day.

At the first faint hint of croup in the voice, respiration or cough, I have seen singularly good results from the application of a sponge dipped in hot water—as hot as the child will bear —directly over the larynx and trachea. The sponge should be well squeezed out, refilled and again applied every two or three minutes, for say half an hour. At the same time, steps should be taken to secure a warm, moist and uniform atmosphere for the little patient, in fact, as Dr. Prosser James urged, many years ago, "the patient should be kept, as it were, in a vapor bath," and of temperature too, much higher than is usual in any sick chamber. The temperature should be kept from 70° to 75°, and I have kept it at 80° for two or three days with excellent results. In order to accomplish this it will be necessary to make a so-called 'croup-tent' around the child's bed, behind that tent a kettle of boiling water is placed on a spirit lamp, and the steam from the kettle is thrown into the tent by means of a long tin spout. But ventilation must be seen to, for fresh air is just as much a necessity as warm, moist air, and without fresh air the croup-tent would be a positive nuisance. I do not recommend the warm bath, believing with Cheyne that "the warm bath is a very equivocal remedy." The diet should be bland and mucilaginous throughout the illness, though well-made beef-tea in small quantities is always in place.

I incline to think that medical men of our school look with a less favorable eye on tracheotomy than do their brethren of the dominant medical faith. The chief cause of this seems to be the great reliance which we justly place upon our therapeutic agents, and hence, Bæhr, our best systematic writer, summing up the resources of the old school against true croup, says: "In spite of all these appliances, from 70 to 90 per cent. of all undoubted cases of membranous croup perish. This result is certainly no triumph, *nor has tracheotomy increased the chances of recovery.*"

Home first suggested tracheotomy in his classic work on croup, in the year 1765, though he never performed the operation. In the year 1782, a London surgeon named John André secured the honor of the first operation, at all events, his is the first recorded case. In 1818 the celebrated Bretonneau revived the operation, but his patient died and the same result followed a second attempt six years later. As an epidemic of diphtheria was raging at the time—the epidemic, in fact, in which Bretonneau won such imperishable laurels—it is highly probable that these were diphtheria patients, hence the fatal result, for in the language of Dr. A. W. Barclay, "tracheotomy is certainly more adapted to this disease (pseudo-membranous croup) than to diphtheria, in so far as the attack is local instead of constitutional, is an inflammation instead of a blood-poisoning." In 1825, Bretonneau again attempted the operation, this time with success, and following his lead, Trousseau, for a time, saved half his patients, though the average of recoveries seems to have been about one-fourth of the whole number of cases.

The older English practitioners had but little confidence in tracheotomy. Cheyne never approved of it, and argued against it with a good deal of skill, and he had very great weight with the English practitioners almost down to our own day. Copland gives a *resumé* of treatment, including bleeding, emetics, purgatives, sudorifics, expectorants, antispasmodics, calomel, blisters and baths, concluding with Valentin's famous recommendation, "the application of the actual cautery upon each side of the throat, in the most severe forms of the disease, when it is at its acmé," to which the distinguished author of the *Dictionary of Practical Medicine* adds, "there does not seem to be a chance from this operation in any case wherein the treatment developed above has failed." Certainly, after the patient has been subjected to the destructive art of healing as exemplified by such a course of treatment, he would be most unlikely to possess sufficient vitality for tracheotomy or any other operation. The older practitioners on this continent sympathized with their English brethren in this matter. Dr. Dewees was very strongly opposed to it; Dr. Physick had no confidence in it, and the operation was little used till a new generation of practitioners arose who leaned less on authority and more on personal experience.

Greve places the mortality in Sweden at 23 per cent.; Trousseau at 50; Franquet at 68; Bricheteau, who draws a sharp line between diphtheritic and true croup, gives the mortality from the latter as 69 per cent. Steiner endorses the operation, yet he says: "The proportion of recoveries is stated by all writers of honesty and diagnostic skill as lamentably small. Out of quite a large number of cases occurring in my practice, *before I had adopted the operation of tracheotomy*, I saw but three recoveries; since 1863, however, this discouraging rate has been so much improved by the employment of tracheotomy that the mortality has, at different times, amounted only to sixty, sixty-five and seventy per cent."

French physicians have employed tracheotomy with marked success, so much so that one feels morally certain that all their patients could not have been blood-poisoned, diphtheritic ones, but that many of them must have suffered from pseudo-membranous croup, and their success arose from the fact that they operated in the early and hopeful stage of the disease, while the English, till very lately, seldom resorted to it till the case was hopeless. Furthermore, I believe that there are cases, not very many it is true, of this disease, *in which tracheotomy offers the only chance of life*—I allude to the *foudroyante* cases, in which the disease marches on to a fatal issue, unchecked by the best selected remedies. In these cases the only safety lies in the prompt and skillful use of the knife.

Tracheotomy does not increase the risk of a fatal issue, and personally I am almost of opinion that the operation is justifiable if only to secure *euthanasia*. The operation has been but little employed by the homœopathic practitioners of this continent, partly because the disease is comparatively rare, and partly because, as already remarked, we have a thorough knowledge of better remedies.

In the first stage of pseudo-membranous croup, I consider *Aconite* beyond all question the leading remedy, for it corresponds not merely to the symptomatic appearances, but it combats the very inmost essence of the disease. In addition to the indications given in the chapter on spasmodic croup, I would add the following, by the venerable and beloved Charles Julius Hempel, who may justly be said to have stamped the peculiar impress of his mind on the homœopathy of this continent: "In *Membranous Laryngitis*, or Croup, Aconite is often sufficient to arrest the inflammatory process which is going on in the lining membrane of the larynx, or to promote its absorption. More than one symptom among the symptoms of Aconite points to its use in croup as a specific remedy. Among the Aconite symptoms we have hoarseness, croaking voices, feeble voice, complete loss of voice, sensitiveness of the larynx to the inspired air as if the mucous membrane were deprived of the epithelium, sensation as if the sides of the larynx were pressed together. These and similar symptoms, together with the dry, hard and tearing cough which Aconite excites, and the raw feeling in the larynx during the paroxysm of cough, are strikingly characteristic indications for the use of Aconite in croup."

Dr. Elb advances the following views: "Aconite, as a medicine corresponding to the local affection and the accompanying fever, must be a perfectly appropriate remedy, and this is corroborated by experience. But as other characteristic symptoms are peculiar to croup, such as the deposition of the exudation, it is evident that Aconite cannot suffice for all cases or stages, and hence that its applicability is limited. Experience teaches us that it is of use when inflammation is still present and accompanied by fever, with hard, full, frequent pulse, and when there is great anxiety and rough respiration. It will accordingly be chiefly suitable for the beginning of the disease, a view in which not only all practitioners are agreed, but this is often laid down as the sole indication." Rückert writes, "Aconite, in fine, should always be administered in the inflammatory stage; it thereby assists the action of the next remedy indicated." Bæhr teaches as follows: "If we are called to a case of croup in the night, it is not always possible to at once obtain the conviction that we are dealing with a case of croup; for even the presence of considerable dyspnœa does not always imply that the disease before us is croup. In order to meet this uncertainty the custom has prevailed for a long time already to at once give Aconite in alternation with some other remedy. We do not approve of this custom of giving remedies in alternation, but make an exception in favor of croup on account of the uncertainty in our diagnosis. Aconite is excellent in catarrhal, but utterly inefficient in membranous croup. If we suspect a case of membranous croup, we give Aconite 2, and Iodium 2, in alternation every hour." Hughes writes, "Whatever medicine you choose, I recommend to alternate it with Aconite. Croup is a neurophlogosis, and the spasmodic paroxysms need as much help as the continuous inflammation." Bæhr and Hughes are undoubtedly, at the present time, the leading homœopathic writers of their respective countries, and yet in opposition to their authority, I would advise Aconite, *and all other remedies*, to be given *singly and alone*. For many years I alternated remedies, and my practice remained destitute of a sound experience. At length, in the Fall of 1869, I made a tour through the Western States, when I noted that nearly all the practitioners who alternated were strongly disposed *to mix medicines*. Finally, the crisis came when a distinguished physician advised me to take equal quantities of Leptandrin 1, Podophyllin 1, and Mercurius solubulis 1, triturate them together, and give this highly-scientific preparation in five-grain doses three times a day as a panacea for "liver complaint." I returned home and never alternated more. Since that time I have adhered unswervingly to the single remedy, and, while I have had vastly better results, I have gradually attained to such an insight into therapeutics as I never could while wandering in the quagmire of alternation. *Give but one remedy.* Should that cease to be indicated select another, but *never alternate.*

Ruddock, following in the wake of Hughes, says: "Even when another medicine is indicated it is often advisable to administer Aconite in alternation to relax the spasm which often complicates the disease."

"A child three years old; severe croup; at the point of suffocating. Aconite 1, one drop in a glass half full of water, a teaspoonful every quarter of an hour. After a few doses profuse perspiration broke out and the child was saved." (Dr. A. Crica.)

"A fat, healthy child, aged two years, was taken suddenly with croup after an exposure to a dry, cold, west wind. Face and skin burning hot; wants to drink constantly; agonized expression; constant restlessness; aggravation after sleeping. Aconite 200, two doses half an hour apart, cured." (Hoyne.)

In this disease I have confined myself to the use of the tincture of the fresh root, or the first and second decimal dilutions of the same preparation, from two to five drops in a tumblerful of water, a teaspoonful every half hour, or even every fifteen minutes. I am aware that I have been censured for recommending the use of mother tinctures and low dilutions, but I would remark that I am not giving the experience of my censors, but my own. Very much more important than an adherence to the high potencies is a thorough knowledge of pathology and pathological anatomy, and a little of the eloquence directed against the low dilutions would not be thrown away if it were turned against the polypharmacy and alternationism and isopathy which threaten to engulf our school.

Iodine is a remedy upon which many physicians rely in this disease, though Kreussler, an excellent therapeutist, says that "he does not recommend it, as our provings upon the healthy do not seem to point to Iodine as a remedy for croup." It was introduced as a remedy for pseudo-membranous croup by Koch in 1841, and since that date Spongia has lost its position and is now only used in the less dangerous catarrhal and spasmodic croups. Koch reports that he gave Iodine alternately with Aconite in thirteen cases of croup, all of which he affirms were pseudo-membranous, with such success that none died; but, as a writer in the *Neues Archiv.* points out, "the result does not speak decidedly enough in favor of Iodine, for a second remedy, often of essential service in croup, was always given in alternation." Still, the same writer admits that in a patient affected with stenosis of the larynx, Iodine produced the most frightful suffocative symptoms and a sound like the most violent croup, and he adds that four cases of the cure of croup by Iodine are recorded by Tietze in the *Neues Archiv.*, vol. I.

Both Elb and Bæhr urge us to give Iodine from the very inception of the disease. Elb writes: "given at the first onslaught of the disease it is calculated to cut short the whole malady;" and Bæhr thinks that "there is no reason why a medicine that embraces in its pathogenetic series all the symptoms of croup, and must therefore be adapted to every stage of this disease, should not be given at the very commencement of the attack." Elb alternates Iodine with Aconite—indeed it seems to be almost impossible to get practitioners to confide in one single remedy in this disease —and even Bæhr, usually a single-remedy man, apologetically says, that "we do not approve of this custom of giving remedies in alternation, but make an exception in favor of croup on account of the uncertainty in our diagnosis." "Like the sudden subsidence of a storm," writes Elb, "so wonderfully quick is the action of this first dose, if the dose was not too strong, the anxiety and imminent suffocation and whistling cough cease, as if by magic, and the dyspnœa becomes so much diminished that we may safely wait an hour before giving a dose of Aconite ; this speedily procures a remission of the fever, with the breaking out of a beneficial perspiration ; the danger is generally past in a few hours, notwithstanding which pause it is not advisable to leave off the medicines too soon, seeing that the disease can only be suppressed and kept down by these means; for which reason I continue the use of Iodine and Aconite alternately every hour, even during sleep, until the breathing is no longer sawing and the cough has become looser, after that only every two or three hours; in this way the transition to an ordinary catarrh is effected, and recovery takes place." Hempel, in his work on Practice, places most reliance on Aconite and Spongia, adding : " If Spongia seems powerless and the spasmodic wheezing continues, we must try Iodine." Hughes considers Iodine "our chief remedy in true croup," and adds that "the medicines between which our choice lies are Iodine, Bromine and Kali bichromicum." Meyhoffer considers that Iodine is most suitable for sporadic croup occurring in previously healthy subjects, when the disease is more sthenic in form, and Ruddock that Iodine should be preferred to Bromine in scrofulous patients.

The symptoms indicating this remedy are not very clearly marked. Koch speaks in general terms of the great value of Iodine in true croup, looking upon it as a kind of panacea, but he gives no special indications for its use. Elb gives the following excellent indications for this " most efficacious and most frequently applicable remedy":

" 1. In cases where there are violent fits of coughing, threatening suffocation, with whistling tone and great anxiety; hissing, sawing, respiratory sound ; painfulness of the larynx; hoarseness and red face, synochal fever; consequently at the first appearance of the disease.

" 2. In cases where there are long continued fits of loose-sounding coughing, without great danger of suffocation, which affords the patient no relief, with slight painfulness of the larynx; strong sawing and hissing but not whistling respiratory sound; temperature of the skin not elevated; with frequent, hard, but not full pulse.

"3. In cases where there is want of cough, or rare, short, loose sounding, but still genuine croupy cough; with constant, but apparently not very troublesome, oppression of the chest, and rough, sawing, not whistling, respiratory sound; cold, moist skin; small, hard, quick pulse.

"4. In cases where the bronchial ramifications are chiefly affected, consequently where there is want of cough, or rare, short cough without the croupy tone; inaudible vesicular inspiration; short, quickened respiration; loss of voice, with weak sawing, rather rattling respiratory sound; abdominal inspiration; painlessness of the larynx and trachea; pale, fallen-in countenance; cold skin, covered with clammy sweat, with weak, small, rapid, and even thready pulse."

There is roughness in the larynx, also painful pressure and stitching in the same organ; pressure in the larynx and pharynx, as if swollen; pain in the larynx with discharge of hardened mucus; constriction and heat in the larynx; increased secretion of mucus in the trachea; dry, short and hacking cough; soreness of the throat and chest when in bed, with wheezing in the throat and drawing pains in the lungs, corresponding with the beat of the heart; great difficulty in breathing; tightness of the chest when breathing deeply; more violent and quicker beat of the heart, with smaller and more rapid pulse; hoarseness, the voice becomes deeper, and finally quite deep; the face is not bluish and bloated, but pale.

Hartmann thinks that tracheal and bronchial croup is the proper sphere for Iodine, especially when there is a tendency to torpor, and he says that the Iodine-croup is always characterized by pain in the chest and larynx. I remember hearing Dr. Constantine Hering make the curious remark, that, while Bromine suited blue-eyed children, Iodine is adapted to black-eyed ones, and other observers have confirmed this. Hempel and Hartmann are at variance as to the precise pathological state, for Hempel says that it is the remedy, "especially in that stage of croup where the exuded lymph begins to become consolidated as an organized membrane, with suffocative wheezing and a fully developed croupy sound during the inspirations," while Hartmann affirms that "there are no symptoms pointing to a pseudo-membranous formation either in the larynx or the upper portion of the trachea." Here I consider that Hempel is undoubtedly correct, for in all my cases, in which Iodine proved curative, false membranes were present. Bæhr gives a few needed words of warning not to be too ready to change the remedy. "In most cases Iodium will undoubtedly have a favorable effect. Only we must not indulge in the expectation of cutting the disease short. A result of this kind only occurs in a very small number of cases. Most commonly the pathological process continues to go on under the action of Iodine, after which it retrogrades, as is the case in every other inflammation. What is essential is that it should be kept confined within proper boundaries. Even if the dyspnœa increases at first, this is no reason why the use of Iodine should be discontinued."

As to the dose, Hempel recommends the mother-tincture; Bæhr the 2nd decimal dilution; Elb the 2nd to the 6th dilutions (centesimal, I presume) and Hartman the 3rd or 4th, going up to the 12th; I have always used the 2nd or 3rd decimal. Whatever preparation is used care must be taken to have it freshly prepared, as a dilution even a few weeks old is not to be depended on. Elb's hint must be kept in mind; by attending to it I have succeeded where success seemed beyond my reach. In all forms of croup it is of importance not to intermit the medicines during sleep, for only by their constant employment is it possible, especially in the bad cases, to stop the progress of the disease." In the year 1858 Dr. William Arnold of Heidelberg introduced the use of Iodine inhalations in cases in which that remedy was indicated, but failed to cure when given in the usual manner. "The evident and visible effect of the Iodine-vapors was looseness of the cough, separation of the membrane, and consequent greater facility of respiration. The mode of preparation was simply to pour a few drops of, from the strong tincture to the second dilution of Iodine into a shallow vessel filled with boiling water. The child was made to inhale the vapor by holding its head over the vessel, or in its immediate neighborhood. The preparation of the vapor was renewed more or less often, as it was needed, from every two to every six hours. At first the vapor appeared to be agreeable to the children, since they endeavored to approach the steaming vessel. Subsequently the effect seemed to be unpleasant, for two of the children resisted the application after the more violent attacks had been relieved." Drs. Hempel, Drake and Schlosser report remarkable success from this simple measure.

To Dr. Allomge belongs the credit of introducing Bromine as a remedy for pseudo-membranous croup, and he asserts that it is the only remedy that can produce the false membrane in the larynx and trachea of the healthy. On the other hand, Dr. Meyhoffer, of Nice, thinks that Bromine is only of use in diphtheritic croup—when diphtheria extends to the air passages; but it has unquestionably proved curative in severe cases of pseudo-membranous croup, as numerous reported cases testify, though Hempel says that it has been used with "variable and rather doubtful success." Hughes says that the specific action of Bromine on croup is unquestionable, and a writer in the *British Journal of Homœopathy*, vol. V, thinks that "we must assign to Bromine the first place among the croup remedies we as yet know." Bæhr is in doubt as to the place and power of this remedy in croup; "instead of Iodine, many physicians recommend Bromine; some successful cures with Bromine are reported, whereas others deny it all power over croup. We are not yet able to express a decided opinion on this subject," and again, "we do not mean to reject Bromine, but it is only in mild cases that we would substitute its use for that of Iodine."

There is spasm of the larynx occasioning suffocation; cough

with croup sound, hoarse, wheezing, fatiguing, not permitting one to utter a word. This cough is generally without expectoration, while with Iodine the cough is generally with expectoration; the respiration is wheezing, alternately slow and suffocative, and hurried and artificial. In this remedy the respiration is with dry sound, while with Iodine the respiration is predominantly with moist sound; the respiration is labored, painful and oppressed, with gasping for air; heat in the face; pulse rather hard, slow at first, and afterwards accelerated. Ruddock advises Bromine "in asthenic croup with extreme congestion and swelling of the air passages, so that the child breathes with great difficulty, throws his head back, grasps at the throat, and evinces anxiety. Affection of the upper part of the air tubes; dry, croupy cough, like that of a sheep, grating and tickling," and Dr. Guernsey's indications for Bromine in croup and laryngeal diphtheria is "rattling of mucus in the windpipe when coughing," which is not very much of a "keynote," after all.

The following appearances are found in the bodies of animals poisoned with Bromine: "Inflammation of the organs of respiration. A quantity of bloody foam in the larynx and trachea. Inflammation in the larynx, trachea and bronchi; sometimes consisting of slight reddish stripes, sometimes of dark redness, sometimes of reddish coloring. Great inflammation of the larynx and trachea, with exudation of plastic lymph, almost completely stopping up the air passages" (*British Journal of Homœopathy*, vol. V).

This remedy has usually been given in the form of dilutions prepared with distilled water—the 1st to the 3d being most highly recommended. The late Dr. E. H. Drake, of Detroit, has used inhalations of Bromine in this disease with eminent success. "My manner of using it is to take a drachm vial about half full of pure water, put in about four or five drops of Bromine—a part only of which will be dissolved, while the residue will fall to the bottom, and be taken up as fast as that already held in solution passes off by its exceeding volatility. Thus the solution may be kept of uniform strength for twenty-four or thirty-six hours. The vial is then held to the *mouth* of the patient, so that the medicine will be inhaled through the mouth, which has seemed to answer better than when inhaled through the nose. The first few inspirations will cause resistance on the part of small children, on account of the unpleasant sensation it produces in the throat; but by letting them take two or three, and waiting a short time, a minute or so, before renewing it, this is easily overcome. Most patients will take it while sleeping. Care should be taken to keep the mouth of the vial well closed with the finger or cork when the patient is not inhaling." Hughes mentions that Dr. Kafka has contributed to the *Allgemeine Homöopathische Zeitung* for 1875, a severe case of membranous croup in which inhalations of Bromine (1st and 2d decimal on cotton wool) had a most beneficial effect.

Kali bichromicum is another remedy which has been used with success, and Dr. Hughes remarks that " there is a large accumulation of evidence tending to show that it is a potent remedy for true membranous croup," and in a later edition he says that it is "of all medicines, most homœopathic to membranous croup—has frequently cured it." Dr. Hempel thinks that it may be used in the last stage when the membrane is formed, and I have seen some remarkable cures with this remedy, especially when the disease extended to the bronchial tubes. Dr. A. E. Small, of Chicago, writes: " I have found this remedy of the greatest use in arresting the progress of membranous croup when the attack comes on in the morning with hoarseness and accumulation of mucus in the larynx, tending to pseudo-membranous formation. After a few doses of Aconite, the 3d decimal, to allay the arterial excitement, Kali bichromicum, the 3d decimal, in water, administered at short intervals, has produced a speedy cure."

The following are the indications for this remedy as given many years ago by the British and Austrian practitioners. The symptoms approach gradually and insidiously; at first, slight difficulty in breathing when the mouth is closed; slight elevation of temperature; pulse irregular and intermittent, or frequent and small; as the disease progresses, the difficulty of breathing increases; the sound of the air as it passes through the trachea is shrill, whistling, as if it passed through a metallic tube; voice hoarse; cough not frequent, but hoarse, dry, barking and *metallic ;* deglutition painful; tonsils and pharynx red, swollen and covered with an appearance of false membrane; after a time, breathing affected in part by the action of the abdominal muscles, and those of the neck and shoulder-blades; head inclined backwards; breath offensive; finally, diminished temperature of the skin; prostration; stupor. Hughes says that the thickness and tenacity of the false membrane will always be an indication for this remedy.

The English provers found the following morbid appearances on the bodies of dogs poisoned with this drug: " Epiglottis and rima glottidis congested and covered with thick, ropy mucus; larynx and bronchi filled with muco-purulent matter. Mucous membrane of larynx, trachea and bronchi deeply injected. Larynx, trachea and bronchi lined with a false membrane, easily detached. In the bronchi polypus-looking masses which could be traced like cords through all the branches of the air tubes." (*British Journal of Homœopathy*, Vol. V.)

As to the dose, Hempel recommends a powder of the 3d trituration dry on the tongue every two or three hours. I have, however, had the best results from a yellow-colored solution of the 2d decimal trituration, giving a teaspoonful every hour or half-hour, and in urgent cases I do not hesitate to give the 1st decimal trituration, similarly prepared.

Many years ago I encountered a very fatal epidemic of pseudo-membranous croup, against which our usual remedies were not as successful as one could wish, while I noticed that allopathic treatment was worse than useless. In my extremity I applied myself to the study of the homœopathic

Materia Medica—that monument of unwearied industry—
and decided that Sanguinaria Canadensis was an appropriate
remedy, as it presented the following symptoms: " *Chronic
dryness in the throat and sensation of swelling in the larynx,
and expectoration of thick mucus. Aphonia, with swelling
in the throat. *Continual severe cough, without expectoration, with pain in the head and circumscribed redness of the
cheeks. Tormenting cough with exhaustion and circumscribed redness of the cheeks. *Croup." In the first
volume of the *Transactions of the American Institute
of Homœopathy* I found that Dr. Bute, the original
prover, considered it to be " very effective in croup."
Soon after, I was called to an undoubted case of
pseudo-membranous croup, and as I had no tincture of
Sanguinaria in my office I gave minute doses of Sanguinarin
in water, and the result was a rapid cure. In the course of
my studies I read Professor Paine's *Epitome of Eclectic
Practice*, in which he gives the following testimony as to the
efficacy of Sanguinaria in this disease : " The Sanguinaria is
one of the most valuable remedies known in the treatment
of pseudo-membranous croup. It has proved as much of a
specific for that disease as Quinine has for ague. I have
seen it used in a great number of cases, and have never
known a single failure, It should be made into an acetic
syrup, by adding twenty grains of Sanguinarin to four
ounces of vinegar; steep and add one ounce of sugar to
form a syrup. Dose, one teaspoonful as often as indicated."
I frequently gave the remedy as Professor Paine directs, but
finding that the large dose caused an unnecessary aggravation, I reduced the quantity, and for a number of years I
have used the following formula : dissolve two grains of the
1st decimal trituration of Sanguinarin in three teaspoonfuls
of good vinegar, adding six teaspoonfuls of brown sugar and
twelve of water, and of this acetous syrup I give a teaspoonful every hour or every half hour. I have given the
Sanguinarin in the 2d decimal trituration, dry on the
tongue, but obtained better results from the acetous

preparation. Of late years I have used the tincture a good deal, especially since reading the following case, reported by Professor Helmuth, of St. Louis, now of New York, in the fourth volume of the *American Observer*:

"We were called to see a child of some ten years of age, who had been suffering from an attack of whooping cough, and who was taken with severe croup. The patient lived about seven miles in the country, and was in a most pitiable condition when we arrived. The previously-existing pertussis had much enfeebled him, and the croupy paroxysms were decidedly the worst which, in fifteen years, we had seen. The suffocative fits were of the most distressing character, and the cough so severe that, as we passed to the house, the noise so resembled the barking of a dog that we were certain that it belonged to the canine rather than to the human species. There were no other symptoms but these: constant croupy cough, excessive suffocation, hoarseness, with tossing about the bed to endeavor to gain air. He had taken for some time *Aconite* and *Spongia*, which had been administered by his parents, without any benefit; also *Hepar sulphur*, with no good result. Having treated him before for severe attacks of this kind, we prescribed *Ipecac* and *Kali bichrom.*, to be taken every fifteen minutes in alternation, and being obliged to return to the city left *Iodine* and *Tartar emetic* to be taken if no relief was experienced after three hours' trial with the previous medicines. This was about twelve o'clock at noon. At midnight we were called again, and after an hour's ride found the little patient in what we then thought a dying condition. The pulse was almost gone, the face livid, the breathing rattling and stertorous, and every symptom indicating a most alarming state of suffocation. The medicines had been faithfully tried, but without result. While the vehicle was being prepared for the ride, we consulted Dr. Hale's *New Remedies*, and reading therein the remarks of Dr. Thomas Nichol, and being really at a loss what to prescribe, we took with us the tincture of *Sanguinaria Canadensis*. Of this were mixed about twenty-five drops in half a glassful of water, and a dessert spoonful administered every ten minutes for half an hour. The symptoms began very gradually to abate, the breathing to become less labored, and the pulse consequently to become fuller. The medicine was continued, at longer intervals, with constant amelioration of the symptoms, and recovery resulted. We are of opinion that had it not been for the work of Dr. Hale this child would *have died*; and *en passant*, would advise our friends to experiment with this valuable remedial agent in similar cases. We are aware many will say, "Where are the symptoms that called for it? What was the 'key-note' that demanded Sanguinaria?" There was *no* key-note but the rattle of death. There were no symptoms but *croup* in its last stages—suffocative breathing and asphyxia.

Tartar emetic is the remedy most generally indicated when paralysis of the pneumo-gastric nerve threatens, though Bæhr thinks it "is indicated if the dyspnœa and danger of suffocation are occasioned by movable patches of membrane. The cough is indeed feeble and without resonance, but a mucous râle is still distinctly heard in the trachea." The respiration is very short, the dyspnœa almost amounts to suffocation, the cough is loose and rattling, a shrill, whistling noise accompanies both expiration and inspiration, the child lacks the strength necessary to expectorate, the chest expands only on the most desperate efforts, and the anxiety and prostration are very marked indeed. The face soon becomes cold and bluish, the forehead is covered with a profuse cold perspiration, and the patient is almost *in extremis*. "Whilst Jahr considers it to be indicated when, after the removal of the dangerous symptoms, much mucous secretion remains, and in the opposite circumstances of a paralyzed state of the lungs. Bosch has recourse to it when the violence of the attack is apparently broken up (transition to the torpid croup?) and others give it only when Hepar and Spongia have been ineffectually employed, without being able to assign a distinct ground for its selection." (Elb.)

My own opinion is that Tartar emetic is a remedy for severe catarrhal croup, but not for the more dangerous pseudo-membranous form. Elb recommends it in the 2d or 3rd dilution, while Bæhr gives grain doses of the 2d trituration and cautions us " not to prescribe this remedy in large doses, for the favorable effect of the act of vomiting is very problematical, whereas the great depression caused by the vomiting is sure to follow."

Phosphorus is recommended by Bæhr if the cough has lost all resonance and force, and the mucous râle has ceased ; or more especially if the croupous process has invaded the bronchia, and the lungs have evidently become hyperæmic, and Elb says that it is most likely to do good when congestion of the lungs and heart with blood is to be regarded as the cause of the pulmonary paralysis. Elb recommends the 2d or 3d dilution ; Bæhr would not dare give it below the third attenuation.

We owe the following indications for Bryonia to Elb, whose classical essay on croup should be in all hands. "The indications for this medicine are completely identical with those of Phosphorus for the remaining cough, only it is to be preferred in those cases where the cough is less deeply seated in the trachea or fever is still present." Dr. Alphonse Teste, many years ago, introduced Bryonia in alternation with Ipecac as remedies for croup. "Ipecac and Bryonia (but given concurrently, for both would be inert alone) are in all cases, whatever be the form of the attack or intensity of the disease, the great modifiers of croupal angina. These medicines need not be prescribed at very low dilutions—from six to twelve will suffice. The two solutions prepared, they should be administered alternately, a teaspoonful every two hours during the period of invasion ; every ten minutes during the exacerbations, and at intervals gradually increased, when these are passed." Jahr says that this was done "by the advice of a clairvoyant," and the late Professor Williamson, of Philadelphia, once told me that it was a communication from the spirit world. Here the Ipecac is prescribed against the spasmodic element of the disease, and the Bryonia against the exudative inflammation, so that this may be looked upon as a model of alternationist reasoning and practice. I used this prescription in former years, and found it effective against spasmodic croup, but utterly useless against pseudo-membranous croup.

Aphorisms.

1. Many of the epidemics of croup during the eighteenth century, would now be styled laryngeal diphtheria.

2. Home and Cheyne remark that the younger children are when weaned, the more liable are they to pseudo-membranous croup.

3. Pseudo-membranous croup is more prevalent among boys than girls, and robust, ruddy, healthful children are most likely to be attacked.

4. Season and temperature exercise a much more powerful influence than constitution and temperament in the causation of this disease.

5. Croup is four times as frequent in the Winter quarter as in the Summer one, and the mean monthly temperature and the mean monthly mortality from croup rise and fall together throughout the entire year.

6. While exposure to cold is the leading exciting cause of pseudo-membranous croup, it must be admitted that in very many cases the exciting cause is absolutely inscrutable.

7. Second attacks of pseudo-membranous croup are very rare.

8. A family predisposition to this disease unquestionably exits, but so far it has not been proved to be hereditary.

9. An obscure epidemic influence is sometimes associated with pseudo-membranous croup, and it is sometimes, but rarely, distinctly endemic, but it is never contagious, though diphtheritic croup is.

10. Croup holds a place intermediate between the zymotic class of diseases and those of the respiratory organs.

11. Pseudo-membranous croup complicates measles, small-pox, scarlatina, whooping cough and typhoid fever, and the larynx is more or less implicated in the majority of cases of acute infectious disease.

12. Croup is probably more influenced by peculiarities of country and climate than any other disease of the respiratory organs.

13. A cold and moist atmosphere, with rapid alterations of temperature, together with the vicinity of the sea, make up the climate in which croup is almost endemic.

14. Hoarseness in a child is of more moment than hoarseness in an adult, and roughness of the voice with hoarse cough should always suggest croup to the mother.

15. In croup, the frequent cough is a better omen than the rare cough, and complete suppression of the cough is one of the worst signs.

16. As a rule, pseudo-membranous croup is marked by remissions, which become shorter as the disease advances, but sometimes it marches directly onward to suffocation.

17. The time for successful treatment is during these remissions.

18. A sudden amendment is more likely to be followed by a relapse than a gradual one.

19. The leading homœopathic remedies are Aconite, Iodine, Bromine, Kali bichromicum and Sanguinaria; minor, but still important ones, are Tartar emetic, Phosphorus and Bryonia.

CHAPTER XI.

DIPHTHERITIC CROUP.

Diphtheritic croup is that most serious variety of croup which results from the development of the characteristic membrane of diphtheria upon the larynx and trachea, accompanied by the blood-poisoning which is part and parcel of the general disease. It calls for a separate essay on account of its very serious nature, and also because many eminent writers and practitioners confidently assert that diphtheritic croup and pseudo-membranous are one and the same disease—that, in fact, pseudo-membranous croup is merely a sporadic laryngeal diphtheria.

In all forms of diphtheria, and, indeed, in every case of the disease, its appearance in the larynx is to be dreaded beyond any other complication. The general opinion is that the diphtheritic membrane reaches the larynx by direct extension from the fauces, but Dr. Wade of Birmingham, England, asserts that he has never found the laryngeal exudation continuous with the pharyngeal. Dr. Ludlum of Chicago holds the opposite view : " The exudation may commence in the larynx or trachea, but is more prone to follow upon that which takes place in the fauces. Sometimes the curtain which envelopes the latter extends through the glottis into the vocal organ, and encroaches upon the trachea, even down to its bifurcation. Such a case would be accompanied by extreme dyspnœa." Oertel of Munich points out that " there are cases on record in which diphtheria localized itself first in the mouth, on the lips, and from these points, skipping the fauces entirely, at once attacked the larynx. Finally, there are rarer ones, in which the diphtheria involved the larynx first, and the mucous membrane of the fauces secondarily, while it also extended downwards into the trachea or bronchi." My own personal experience is, that as a very general rule, the disease originates by extension from the fauces, and that it is a very rare thing to find it originate in the larynx. I remember one notable case in which, when I first saw the patient, the only diphtheritic membrane to be found covered both eyes like two patches of very thick cream ; no membrane in the fauces and no laryngeal symptoms whatever. In twenty-four hours the larynx became inflamed, abundant diphtheritic membranes were thrown out, and the patient died forty-eight hours after the first appearance of the laryngeal symptoms. At death, no diphtheritic membrane was to be found in the fauces.

The larynx is likely to be affected when the diphtheritic membrane covers the fauces very completely, extending very far down into the pharynx, though I have seen a number of cases in which large patches of diphtheritic membrane covered both tonsils, the mucous membrane of the pharynx being almost normal, when suddenly the dreaded laryngeal symptoms appeared, and that, too, at a very early period of the disease. This was possibly caused by multiple infection, but more likely by breathing the poison contained in the mouth and fauces—a true secondary infection.

Physicians well read in historical medicine know that diphtheria is not, by any means, a new disease, but an old disease which manifests itself only at somewhat long intervals, and also that all writers on diphtheria make mention of diphtheritic croup. Aretæus, the Cappadocian, styled by Squire " the founder both of our knowledge and treatment of diphtheria," mentions that the disease, styled by him *Egyptian and Syrian Ulcer*, sometimes extended from the fauces to the windpipe, where it proved rapidly fatal by suffocation, and he adds that children under puberty are especially subject to the laryngeal complication. The Spanish physicians Herrera, Villa Real and Fontecha, who wrote in the beginning of the seventeenth century, give in their works most excellent descriptions of the disease which we call diphtheria, styled by them *garrotilla* or *morbus suffocans*, on account of the laryngeal complication. Alaymus, who describes the diphtheria epidemics of Sicily in the early part of the seventeenth century, speaks of the disease as extending to the larynx, and also of its commencement there. In the year 1753, Dr. Cadwallader Colden, of New York, observed a disease which could only be diphtheria. " It is attended with a moist, putrid heat, the skin being seldom parched. The pulse is usually low, but frequent and irregular. The countenance dejected, with lowness of spirits; no considerable thirst; the tongue much furred, and the furring sometimes extends over the tonsils as far as the eye can reach. At other times, in the mildest kind, the tonsils appear only swelled, with white specks of about a quarter of an inch or half an inch in diameter, which are thrown off from time to time in tough cream-colored sloughs. Sometimes all the parts near the gullet or throat are much swelled, both inwardly and outwardly, so as to endanger suffocation, and frequently mortify; but most generally the swelling internally is not so much as to make swallowing difficult. Sometimes those swellings imposthumate. The last complaint is commonly of an oppression or strictness in the upper part of the chest, with difficulty of breathing, and a deep, hoarse, hollow cough, ending in a livid, strangled-like countenance, which was soon followed by death." From the last sentence we are led to conclude that croup, undoubtedly diphtheritic in its nature, was the usual termination of the disease, and the same peculiarity has been observed in other epidemics.

In the year 1755, Dr. Richard Russell, of London, noted croup in connection with an epidemic of malignant angina, and there can be little doubt but that in our day that this malignant angina would be styled diphtheria, and that the accompanying croup was diphtheritic in its nature. Ten years later Francis Home wrote his classic work in which he describes what we now style pseudo-membranous croup, and he was very careful to draw the diagnostic lines between it and diphtheritic croup, which he had certainly seen. In 1779, Dr. Johnstone, of Kidderminster, insisted on the essential difference between the two diseases, but, in the words of Dr. Squire, "Unfortunately, though argued with learning and experience, these views did not prevail; the name of croup was applied to the epidemic complication, and the treatment laid down by Home for the one disease was very energetically employed against the other." The epidemics of croup at Cremona and Liskeard, referred to in the chapter on pseudo-membranous croup, appear to have been closely akin to the disease described by Dr. Richard Russell, and the latter half of the eighteenth century and the first years of the nineteenth are notable for numerous outbreaks of malignant angina associated with laryngeal inflammation. The "Suffocative Angina," so well described by Dr. Samuel Bard, of New York, in 1771, appears to have been identical with diphtheria, and he remarks that he saw upon several children in the same family thick, coriaceous pellicles formed upon the tonsils, and propagated from the pharynx to the trachea. "Three *post-mortem* examinations exhibited to him, as a uniform result, white, thick, coriaceous, elastic layers of concrete matter, which lined the walls of the pharynx. A membraniform tube of the same nature advanced into the trachea and became progressively thinner in proportion as it descended into the bronchi. The tracheal mucous membrane was slightly inflamed; that of the pharynx, after the pellicles were removed, was found rather pale." Dr. Squire claims that the disease seen by Dr. Bard was diphtheria, and certainly the description of the laryngeal complication exactly tallies with diphtheritic croup, while it is quite unlike the morbid state which we call pseudo-membranous croup. To quote Dr. Squire: "Epidemic croup is strictly diphtheria; when that disease prevailed epidemically in England at the end of the last century, any fresh outbreak was so spoken of; an outbreak at Chesham, in Buckinghamshire, in 1793, carefully described by Mr. Rumsey, leaves no doubt upon this point; sometimes on its appearance in a fresh locality it was simply called croup, and the word excited as much terror then as diphtheria has again given us reason to associate with the disease it now designates."

Many of the essays presented to the great Parisian *concours* on croup (1807) really describe diphtheritic croup, and confusion reigned till in 1818 the illustrious Bretonneau investigated the epidemic of diphtheria at Tours. The conclusions at which he arrived have influenced the views of French physicians down to the present day, and, as is well known, the vast majority of them agree in considering that diphtheria and true croup are one disease, the latter being a mere local manifestation of the former. A goodly number of practitioners, among them Bricheteau, Emangard and Desruelles, opposed these views, and the original work of Dr. Bland, of Beaucaire, entitled *Nouvelles Recherches sur la Laryngo-Tracheite* is even to-day one of our best authorities on the subject. In 1820, Dr. Mackenzie, of Glasgow, described a disease strikingly similar to that seen by Bretonneau, and from that date the two diseases were described separately. In 1826, Dr. Abercrombie described a fatal throat affection extending to the windpipe, which was very fatal among the children in Edinburgh, and he adds that "it is evidently quite distinct from the idiopathic inflammation of the mucous membrane of the larynx to which we commonly apply the name of croup." During the succeeding decades, a malignant angina, unquestionably diphtheritic in in its nature, often accompanied by a laryngeal complication, raged at intervals in the three Kingdoms, and the laryngeal complication was always carefully distinguished from true croup.

In 1858, the great epidemic of diphtheria invaded England, and though diphtheritic croup was at first considered to be a distinct affection from pseudo-membranous croup, Dr. Prosser James, in the first edition of work, *Sore Throat, its Nature, Varieties and Treatment*, contended that " both are alike the manifestation of an inflammatory condition tending to exudation"—exudation being, in fact, regarded as the grand characteristic of the disease. Dr. R. H. Semple, who had probably been inoculated with the French views while translating Bretonneau's *Memoirs* for the *New Sydenham Society*, had long held that the two diseases are really one, and towards the close of the epidemic (1868) Dr. Thomas Hillier, a distinguished writer on the diseases of children, declared that he could detect no difference between membranous croup and laryngeal diphtheria. A few years later Dr. George Johnston and Dr. Morell Mackenzie, the greatest authority on throat diseases in Great Britain, gave in their adhesion to the new doctrine, and in 1870, Sir William Jenner, who had long held that the two diseases were distinct entities, finally pronounced himself "inclined to think that the two diseases are really identical." Sir William enunciated his new views with great force and eloquence at a debate on the subject at a meeting of the Royal Medico-Chirurgical Society. The discussion originated in the report of a committee recommending that the word "croup be henceforth used wholly as a clinical definition, implying laryngeal obstruction, occurring with febrile symptoms in children"—a doctrine which, if acted upon, would turn the pathology of the subject backward two centuries.

In the United States the greatest names in pathology—George B. Wood, Austin Flint, J. Lewis Smith, Fordyce Barker, Henry Hartshorne and others—hold that true croup—pseudo-membranous croup—is a wholly distinct disease from laryngeal diphtheria, and I can only recall two names of note, J. F. Meigs and A. Jacobi, who hold the contrary view. Dr. Jacobi attributes this result to the influence of the writings of Vogel and von Niemeyer upon the American medical mind, and certain it is that the views of Meigs and Jacobi are held by but a small majority of the medical men of this continent.

In Germany, many eminent pathologists, among them von Niemeyer, Oppolzer, Letzerich, Vogel, with Rudolph

Virchow, the most eminent of them all, deny the unity of diphtheria and pseudo-membranous croup. Steiner talks on both sides of the question, stating that "the attempt to distinguish croup and diphtheria as two entirely distinct diseases has been unsuccessful, both from an anatomical and from a clinical standpoint." Yet he concedes that "in diphtheria the lesion is similar to that of croup, only with this difference, that in croup the exudation takes place *upon* the free surface of the mucous membrane, while in diphtheria it occurs at the *within* the tissue, and thus produces necrosis and loss of substance of the mucous membrane." This is an important distinction from the anatomical standpoint, and he finally admits that while "true croup is not a contagious disease," "diphtheritic croup possesses this quality in a marked degree," which is certainly of great moment from a clinical standpoint.

In some epidemics the appearance of the diphtheritic membrane on the larynx and trachea is of common occurrence, while in other epidemics it is very rare. Why it is so we cannot tell, but though unexplained, still the fact remains a fact. In low, swampy land, and on the banks of lakes and ponds, laryngeal diphtheria is far more common than on high, rolling land. While practicing in Simcoe, Province of Ontario, I found that diphtheritic croup was more apt to appear near Lake Erie than at a distance from it, and that a river or creek had not the same deleterious effect as bodies of standing water. So common was laryngeal diphtheria in some of the Spanish epidemics of the seventeenth century, that the entire morbid state was styled *morbus strangulatorius* or *garrotilla*. It was very common in the famous epidemic of Tours, so well described by Bretonneau. "In comparing together the morbid lesions observed in fifty-five subjects of all ages, who, in the course of two years had fallen victims to epidemic angina, I find that it once happened that the false membrane existed in the trachea without any exudations being found either upon the tonsils or upon any other part of the pharynx. Six or seven times, that is to say, in the proportion of one to nine, the membraniform exudation reached to the extreme ramifications of the bronchi. In a third of the whole number it passed beyond the great division; in all the rest it terminated at different distances from the trachea, so that the mechanical obstacle offered to respiration by the development of the false membrane always appeared to be the immediate cause of death. A single exception was observed. A child who appeared to die of exhaustion, on the fifteenth day, from malignant angina, without any other symptoms than a continuous vomiting, had the pharynx lined with thick pellicle, which did not pass beyond either the commencement of the œsophagus or the entrance of the glottis" (*Bretonneau's First Memoirs*, 1821). Mr. Thompson, of Launceston, England, says that of 485 cases which came under his observation, the air passages were involved in only fifteen, eleven of them dying within a few hours of the commencement of the croupous breathing. Mr. Schofield, of Highgate, near Birmingham, had thirteen fatal cases of diphtheria in his practice, in three of which it assumed the form of croup. Dr. J. F. Meigs, of Philadelphia, lost six patients with diphtheria, and "in all but one the fatal termination was caused by the extension of the exudation to the larynx." Dr. Capron, of Guilford, England, had nine fatal cases of diphtheria, three of them dying croupous. Dr. Heslop, of Birmingham, England, thinks that the disease attacked the larynx in about five per cent. of the cases he had seen in that city. Of 26 fatal cases of diphtheria reported by correspondents of the *British Medical Journal* (1859) nine, including one from bronchitis, died from the laryngeal complication. Dr. Squire remarks that two-thirds of his cases of diphtheria suffered from laryngeal complications, and that the mortality was very high, about 80 per cent. of the croupous cases. This closely corresponds with the statistics of M. Roger, for the Children's Hospital in Paris, in 1859 and 1860, which show a mortality of about 77 per cent. in laryngeal diphtheria. Dr. Crichton, of Edinburgh, gives the results of 45 cases of diphtheria observed in his practice. Of these, 25 were males and 20 females; nine proved fatal, or 20 per cent. Six of the deaths were from extension of the diphtheria to the larynx, and the remaining three died of asthenia. The average age of the fatal cases was seven years. Dr. Hillier says that "of the cases of diphtheria that have occurred in the Children's Hospital (London) two-thirds of the cases have suffered from laryngeal complications." Meigs and Pepper remark that "the frequency of its occurrence varies much in different epidemics, the proportion varying from one or two per cent. to as high as fifty per cent. of all the cases." Oertel thinks that "the younger the patient the greater is the danger that even the lighter forms of the disease may involve the larynx, while the more extensive inflammations take this dangerous course almost invariably." Jacobi, whose immense experience entitles him to the most respectful consideration, says, "I do not know that sex exerts any predisposing influence over diphtheria, yet of the 600 cases or thereabouts of laryngeal diphtheria in which I either personally performed tracheotomy, or observed the progress of the disease in the practice of others, I found the majority in males, and the recoveries in inverse proportion to the number thereof; the mortality being greater among boys." The writer has seen over eighteen (18) hundred cases of true diphtheria, besides many hundreds of cases of pseudo-diphtheria—a form of morbid action which may be said to bear the same relation to true diphtheria that cholera-morbus does to Asiatic cholera—and the result of his observation is, that laryngeal symptoms have appeared in eighteen per cent. of all the cases of true diphtheria seen from 1858 to 1870, and only three per cent. in all cases of true diphtheria seen from 1871 to 1884. I have never observed laryngeal complications in all the cases of pseudo-diphtheria. Of those attacked with diphtheritic croup, a very large proportion died—not less than seventy per cent.—and the fatality was largely influenced by the locality.

Diphtheritic croup, then, appears in two forms, as an idiopathic affection and as an extension of the disease from the fauces; the first form is quite rare, the second is the most common phase of the malady. When it appears as an idiopathic disease the local symptoms are commonly preceded for some days by slight fever, which is, in the early stage, much less severe than in the case of pseudo-membranous croup; and from the very faintest inception of the morbid process, an amount of depression is present which is out of all proportion to the local symptoms, for the very source of life is already being prostrated by what some English writers oddly term a "morbid poison." It must be noted that as soon as the exudative inflammation attacks the larynx, the fever rises at once, and, curiously enough, the feeling of prostration seems to pass away to a considerable extent. Jacobi accurately remarks that "fever and pain are not necessarily prominent symptoms," and in some of my worst cases the patients had hardly any fever and made no complaint of pain. Soon a slight cough, not at all hoarse, comes on, and this is preceded for as much as twenty-four hours by a slight trilling sound in the larynx, only to be detected by ausculation, which should constantly be used in all cases of diphtheria. But this slight and apparently trifling cough speedily assumes the loud, clangorous character of a true croupous cough, and, at the same time, the respiration becomes stridulous. The cough, which evidently causes great suffering to the child, has, in the graphic words of Oertel, "a peculiar, barking, flat sound without resonance," which an experienced ear can readily differentiate from the cough of pseudo-membranous croup. The roughness and hoarseness of the voice increases with alarming rapidity, and soon complete aphonia sets in, though the mere act of speaking seems to cause no pain. Inspiration is very slow, long-drawn and whistling, while expiration is short and superficial. Suddenly, a frightful paroxysm of dyspnœa sets in, apparently caused by spasm of the laryngeal muscles. Suffocation now appears to be imminent; the child cannot lie down ; the pale, bluish skin is covered with perspiration, and all the powers of life fail rapidly. I have known death occur during one of these paroxysms, and this is apt to be the case when a feeble child has been thoroughly saturated with the diphtheritic poison. As a general thing, however, the paroxysm passes away and is replaced by an interval of comparative ease and restfulness, but soon another paroxysm comes on with still more alarming phenomena, leaving the child still more exhausted. The intervals between the paroxysms become more and more brief till the child becomes comatose and passes quietly away.

Another group of cases presents very similar symptoms, but without the distressing paroxysms of suffocation. In these the pseudo-membrane forms and thickens, the cough becomes more frequent and more severe, and the disease is quite similar to true croup, but with less febrile reaction, and with the peculiar cough already described. As the larynx becomes more and more blocked up with the diphtheritic membrane, the respiration is quickened and the dyspnœa finally becomes extreme ; the inspiration is whistling and very much protracted; the face is pale and anxious ; the sufferer vainly seeks for relief from change of posture, and finally death ensues, though, as Dr. Charles West accurately remarks, " without being ushered in by that urgent dyspnœa and those violent efforts to obtain air which attend most cases of cynanche trachealis." The cough is weaker and less frequent as death approaches.

In the second class of cases the patients are attacked with diphtheritic croup in the course of the ordinary pharyngeal diphtheria. It may either come on suddenly, or the extension may be marked by a little huskiness and weakness of the voice, while the breathing is irregular and imperfect. Soon the well-known cough comes on with extreme dyspnœa, sopor supervenes, and death closes the scene often within a few hours of the laryngeal attack. Dr. J. Lewis Smith says that " occasionally, by great effort on the part of the child or by fortunate treatment, a portion of the pseudo-membrane is expectorated, and for some hours there is apparently great improvement, but it is only in exceptional cases that the plastic formation is not speedily and fully reproduced."

In patients in whom the disease has extended from the fauces to the larynx, the glands of the neck and throat swell, the tongue gets very red, especially at the tip, and a thick, yellowish fur of foul smell covers the entire organ. The breath is extremely offensive, and a thin and fetid fluid runs from the nose and eyes, and, if the patient lives long enough, false membranes form on both nose and eyes. The urine is scanty and high-colored, and albuminuria appears about the fifth or sixth day. At times the urine is wholly suppressed, and I have never known these cases to recover. Oertel and Henoch think that diphtheria of the trachea rarely occurs without the co-existence of diphtheria in the fauces, but I have repeatedly remarked it, but not of late years.

In patients in whom the disease is primarily developed in the larynx, precisely the same group of symptoms makes its appearance, but not so virulent, unless, indeed, the patient should live a week or longer after the appearance of the laryngeal diphtheria.

When a favorable termination is about to take place, improvement may be looked for about the third or fourth day. One of the earliest of the favorable signs is the increased facility of swallowing, and this may take place even when the pharynx and its diphtheritic membrane is apparently unchanged. Next, membranous shreds are expectorated, or still more frequently they are swallowed. Sometimes such a mass of them passes into the intestinal canal that the patient is made quite sick, and this is one of the very few instances in which the homœopathic physician is justified in administering a laxative. On examining the stools, large quantities of the characteristic membrane are found, and the patient soon brightens up on being relieved of the offending substance. *Tolle causam.* The fever decreases, and is succeeded by long-continued sweats; the alarming laryngeal symptoms decline, though hoarseness the result of a partial paralysis of the vocal cords together with weakness of the laryngeal muscles, lasts for quite a time after restoration to health; nosebleed comes on without any assignable cause; and the quantity of urine is greatly increased, while, at the same time, the albumen finally disappears.

During the process of cure, important changes take place in the diphtheritic membrane itself. We see these changes in the pharynx, and we infer that similar changes take place in the laryngeal membrane. Dr. D. Francis Condie gives the following excellent account of these changes: "In favorable cases, as the membranous exudation becomes detached its place is quickly supplied by a new formation, and after each separation it becomes, in general, white, and much thinner. In other cases, the exudation, instead of being separated in fragments, becomes, in part, softened to a pulpy consistence, and is discharged from the mouth mixed with bloody mucus. This separation and renewal of the pseudo-membranous deposit continue, in most cases, for the space of eight or ten days. When, finally, it ceases to appear, it leaves, most generally, the mucous membrane to which it has been attached perfectly sound throughout its whole extent; of a light-red, uniform color, and covered, usually, with a thick, yellow mucus, more or less resembling pus. At the same time the aspect of the child is greatly improved. The features brighten and lose the dull and haggard look which characterizes all serious diphtheria; the tongue becomes moist and clean; the skin warmer, moister and more natural, and slowly, very slowly, the patient regains his former health and strength."

On the other hand, should the case be about to terminate unfavorably, the disease marches on with steady and rapid strides; the respiration becomes more and more stertorous; the cough becomes weaker, and finally is entirely suppressed; the face becomes livid and ghastly; the skin cool and of a dull, dusky, purple hue; the child sinks into a partially comatose state, and death often takes place, as Dr. Ludlam remarks, "without a sigh or a groan."

Dr. Thomas Hillier remarks that "the laryngeal symptoms set in on the first, second or third day; in a very large proportion of cases within the first week," and he adds that he has "seen them occur once on the twelfth and once on the nineteenth day of the disease." Dr. J. F. Meigs says: "If it extend into the air passages very soon after the invasion, it may cause death within a few days. In most of the cases, however, the larynx does not become implicated under five or six days. In one of my cases death occurred on the fourth day, in one on the fifth, in one on the sixth, in two on the seventh, and in one on the eighth." Dr. J. Lewis Smith gloomily and yet accurately remarks that "when the croupy cough, voice and respiration are observed, he will seldom err who predicts a fatal result within a week, and often death follows in two or three days." Oertel's experience is "that diphtheria of the larynx and lower air passages in children usually runs its course in a few days; in from two to eight days, or more rarely as late as the tenth or twelfth day, either a fatal termination or convalescence takes place."

Dr. Charles West gives the following interesting table of 27 cases of diphtheria in which death took place chiefly from the affection of the larynx:

The child died on the	2nd day in	1.
" " " "	3rd "	4.
" " " "	4th "	1.
" " " "	5th "	4.
" " " "	6th "	3.
" " " "	7th "	1.
" " " "	8th "	1.
" " " "	9th "	1.
" " " "	10th "	1.
" " " "	11th "	1.
" " " "	12th "	1.
" " " "	13th "	3.
" " " "	14th "	1.
" " " "	15th "	2.
" " " "	21st "	1.
" " " "	23rd "	1.

My own experience is very similar to that of the

distinguished authors just quoted. Much depends on the malignancy of the general disease, but when diphtheritic croup is really developed, death usually takes place within a week of the invasion of the laryngeal symptoms. Indeed, it may be said to be a general rule that when death takes place in diphtheria within a week, it is by extension of the disease to the larynx. Very few croupous cases live to see the commencement of the second week, for, when death takes place after the expiration of the first week, it is very frequently caused by exhaustion. It is true that I have seen the disease developed as late as Dr. Hillier has observed, but then it could always be traced to an accidental exposure to cold.

In diphtheritic croup death may be the result of a severe and long-continued spasm of the glottis, which seems to be of the essence of the paroxysms already described, or it may arise from a purely mechanical blocking up of the larynx, trachea or bronchi by the diphtheritic membrane, or, lastly, death may, according to Oertel, result from insufficient decarbonization of the blood, due to its unequal distribution;" this inequality in the distribution of the blood is due to the fact that emphysema and anæmia have established themselves in the parts open to the circulation of air, while in the collapsed parts, to which the air does not have access, there is hyperæmia. Later in the course of the disease, pneumonia may set in, or pulmonary œdema, and if the patient should surmount these manifold dangers there remains the blood-poisoning of the primitive disease, which may prove fatal even months after all danger from the respiratory organs has passed away. As a general rule, it is rare to find one of these causes of death acting singly and alone, for the diphtheritic blood-poisoning is an almost invariable factor in the fatal result.

If the throat is examined in the early stages, the fauces, soft palate and tonsils will either be found to be of an universal purplish red, or marked and blotched with the same hue. This redness is succeeded by a thick, albuminous lymph, which is more abundant at the base of the arch of the palate than above it, looking as if it had extended from the larynx. The contrary is the case when the larynx is secondarily affected, for then the lymph is more copious at the summit of the arch of the palate.

Dr. Paul Guttmann, of Berlin, observes that " Very young children, who are most frequently attacked by laryngeal diphtheritis, can very seldom be subjected to laryngoscopic examination ; the affection, however, can usually be diagnosed without it, as we know from experience that symptoms of stenosis of the glottis (crowing and prolonged inspiration), and hoarseness or aphonia, when they present themselves along with diphtheritis of the pharynx, are always due to an extension of the disease to the larynx. Even when the pharyngeal diphtheritis is wanting the above-mentioned indications, when observed in young children in districts in which diphtheritis is prevalent in an epidemic form, generally warrant one in assuming confidently the diphtheritic nature of the laryngeal affection." Oertel gives the following sketch of the results of the laryngoscopic examination made in children sick with laryngeal diphtheria : "All the parts of the larynx will be found intensely reddened and swollen, the epiglottis thickened to twice its natural size, and the yellow colour of the cartilage, which normally shows through its covering, no longer distinguishable ; the aryteno-epiglottidean folds, the false and true cords, are greatly swollen, and are covered, more or less, with a grayish-white exudation, or the interior of the larynx itself is lined with a white, leather-like covering, and the glottis is narrowed. Tenacious exudation and purulent mucus, which push up from the deeper parts of the air-passages, often adhere between the vocal cords, and are driven up and down in the narrow cleft by the forced respiration."

It is well to remember that, even when the lungs are not directly implicated, the respiratory murmur is so overwhelmed by the loud, laryngeal sounds that no vesicular murmur can be detected by auscultation.

The pseudo-membranes found in the larynx vary much in consistence, extent and appearance. Sometimes it is soft, gelatinous and almost liquid, lying loose in the throat, almost like a clot of cream, with its particles so soft and so little connected with each other that it almost seems a misnomer to apply the term 'pseudo-membrane' to it. At other times it resembles a fragment of moist kid glove leather—dense, elastic, coherent, and as thick as a silver half-dollar. I remember examining one pseudo-membrane, in 1860, the upper part of which was a quarter of an inch in thickness, and of horny hardness. It was lying almost loose in the fauces, and on being removed, it was perfectly reproduced in twenty hours. Between these extremes one meets with membranes of great variety as regards thickness, cohesion and appearance. The more liquid membranes are whitish or yellowish-white in color, while the denser ones are grayish, or ash-coloured, and sometimes brown or blackish.

On examining, after death, the bodies of those who have died of diphtheria of the air passages, the epiglottis is often found to be so enormously swollen as to close the entrance to the windpipe. It is also covered on both sides, or on one only, with the characteristic exudation, on removing which, small points of ulceration are found studding the surface. This exudation is not an effusion *upon* the free surface of the mucous membrane, but an exudation *within* the tissue as well, and it often destroys both mucous membrane and epithelium. The larynx is lined with a similar pseudo-membrane, generally whiter than the diphtheritic membrane found on the tonsils and pharynx, on removing which, an ulcerous surface, raw and sore, is seen, for diphtheria—at least in its local manifestations in the fauces and air passages—is simply a specific inflammation with partial necrosis and sloughing of the mucous membrane. The diphtheritic membrane is not so easily pulled off as the pseudo-membrane of true croup, for in the latter disease an effusion between the mucous membrane and the pseudo-membrane is an effort of nature to cure the disease. But in diphtheritic croup the mucous membrane dies, if the case lasts any time.

Dr. Charles West remarks that he has "in no instance observed false membranes extending below the larynx," but, in common with many other physicians, I have seen complete casts of the trachea, bronchi and bronchial tubes, even to the third bifurcation. In the upper part of the trachea the pseudo-membrane is similar to that lining the larynx, but as it descends it becomes thinner and less consistent, till it tapers off to a very thin and transparent pellicle. In some ten or twelve cases I have noticed blood that had been accidentally drawn, and it was always of a dark brownish hue and deficient in coagulability.

Is pseudo-membranous croup identical with diphtheritic croup? Is each case of pseudo-membranous croup merely a sporadic case of laryngeal diphtheria? This is one of the burning questions in pathology, and yet it has been strangely ignored by almost all the authors of our school who have written on diphtheria or on croup. The question at issue is thus temperately stated by Drs. Meigs and Pepper: "But further, our personal experience constrains us to state that the differences between the two forms of membranous-croup above enumerated have not seemed to us sufficiently marked and constant to positively establish their essential diversity; and that it is our decided opinion that the vast majority of the cases usually termed pseudo-membranous laryngitis (pseudo-membranous croup) are, in reality, instances of primary laryngeal diphtheria (diphtheritic croup), in which the constitutional symptoms are not grave, and where the faucial deposit has been very slight and perhaps even overlooked."

As already remarked, this view is held by the great mass of French pathologists, a large number of the Germans, and a small but respectable minority of the English, while on this continent the contrary view is very generally maintained. Another view is taken of this deeply interesting and important subject by a large body of medical men, who look upon diphtheria, in the words of Dr. Charles West, as being "a second form of disease, resembling croup in some respects, though differing in others, alike but not the same." This is the view held by the present writer, who considers that while the distinction between the two maladies is sufficiently marked in typical cases, that in a small number of instances there is a tendency in one disease to run into the other, precisely as every experienced practitioner has met with cases of disease which taxed his diagnostic skill to decide whether they were small-pox or chicken-pox. Further, it is freely conceded that diphtheria implicating the air-passages, must, in the very nature of things, produce symptoms strikingly similar to those of croup, but that by no means proves their identity, for a foreign body in the larynx simulates croup very closely. But, as a very general rule, the non-epidemic, non-contagious, sthenic, localized inflammation of true croup is readily distinguished from the epidemic, contagious, asthenic, general disease which we style diphtheritic croup, in which the local inflammation is merely one of the many incidents of a deeply-pervading constitutional affection.

Dr. Jacobi of New York propounds the following queries: " Can pseudo-membranous croup be distinguished from laryngeal diphtheria? Ought these terms to be preserved separately? Are they different processes? Let us suppose two cases of membranous impediment in the larynx, the one with, the other without membrane in the pharynx, the other symptoms being the same, is one " diphtheria of the larynx," and the other " croup"? Suppose again, a membranous stenosis of the larynx, to which is only later added a membrane of the pharynx, was the case originally one of " croup" which became a " diphtheria" later on? Thirdly, take two cases of laryngeal stenosis, one with symptoms of suffocation only, the other having these symptoms together with adynamia; is the latter " diphtheria" alone, the former only " croup"? In my opinion, it is just as little possible to differentiate these diseases according to the seat of the morbid product, as it is justifiable to deny the title diphtheria to membranous pharyngitis when few general symptoms, such as fever, debility and collapse, happen to be present." To these queries Rokitansky's definitions are a sufficient reply: "Croup is a fibrinous exudation effused in a liquid form, and coagulating on the surface of the mucous membrane, this being unaltered or nearly so," and, "Diphtheria is a necrotic process, consisting in infiltration of the mucous membrane, accompanied by exudation and followed by sloughing," and the chief point of resemblance is that both are manifestations of an inflammatory condition tending to exudation. As Herschel clearly points out, croup is a plastic disease, while diphtheria is a gangrenous one, and this *dictum* holds good, even though the diagnosis of some few cases baffles the most experienced.

Hirschel points out that while in diphtheria the sub-mucous tissue is affected, besides the mucous membrane, in croup only the mucous membrane is the seat of the disease, while L. Fleischmann considers that the membrane of croup is a true pseudo-membrane lying on the surface of the mucous-membrane, from which it can be removed, while the membrane of diphtheria is never a true croup-membrane, but deposits consisting of degenerated and exfoliated epithelium, fungi and detritus. Dr. T. H. Green observes, "It is difficult in many cases to draw any line of demarcation between the histological changes occurring in diphtheria and those of croup. In diphtheria, however, the sub-mucous tissue usually becomes more extensively involved, so that the false membrane is much less readily removed The circulation also becomes so much interfered with that portions of the tissue lose their vitality, and large ash-colored sloughs are formed, which, after removal, leave a considerable loss of substance." Again, on removing the membrane of croup, the mucous membrane remains smooth and uninjured, while in diphtheritic croup, on removing the exudation, the surface is ulcerated and gangrenous. Dr. A. W. Barclay, however, remarks that "the fibrinous exudation, so unusual in inflammation of mucous membranes, is also apparently identical; but as far as we know, the cause is different."

Dr. Morell Mackenzie ridicules the idea of supposing "that there are two kinds of pellicular inflammation of the larynx, one in which the cause is the diphtheritic poison, and the other in which the cause is some other undiscovered influence, is totally opposed to all probabilities;" but, as was pointed out by the croup-diphtheria Committee of the Royal Medico-Chirurgical Society, "Membranous inflammation, confined to or chiefly affecting the larynx or trachea, may arise from a variety of causes, as follows: (*a.*) From the diphtheritic contagion; (*b.*) by means of foul water, of foul air, or other agents, such as are commonly concerned in the generation or transmission of zymotic diseases; (*c.*) as an accompaniment of measles, scarlatina or typhoid, independently of any ascertainable exposure to the especial diphtheritic infection; (*d.*) it is stated, on apparently conclusive evidence, that membranous inflammation of the larynx and trachea may be produced by various accidental sources of irritation—the inhalation of hot water or steam, the contact of acids, the pressure of a foreign body in the larynx, and a cut throat."

The mode of death in the two diseases is strikingly different, though, as a matter of course, the termination of all cases when death results from apnœa is identical. "In croup," as Dr. Prosser James points out, "the exudation endangers life, both by inducing spasmodic closure of the glottis and by mechanically impeding the entrance of air into the lungs; the patient dies suffocated; in diphtheria it is associated with intense depression of the vital powers, such as we see in malignant fevers, and speaks plainly of blood-poisoning; the patient dies exhausted."

Dr. Alfred Meadows, of London, concludes his essay on the identity of the two diseases by remarking, "*At any rate, we must admit that they both are blood diseases*"—which we concede without demur. Scarlatina and syphilis are both "blood diseases," but few would draw arguments in favor of the identity of these diseases from that fact.

Another point of difference is the site of the disease

Diphtheritic croup, in the vast majority of cases, commences in the pharynx and extends thence to the air passages, while pseudo-membranous croup commences in the larynx, and, if it spreads at all, it extends to the trachea and bronchial tubes. In croup the earliest symptom is that of stridulous voice and respiration; in diphtheria the uneasiness is first felt in the fauces. Dr. Hauner, of Munich, says that "true croup always commences in the larynx"—it would be more correct to say that it *generally* commences in the larynx, while diphtheritic croup *generally* commences in the pharynx.

Dr. Morell Mackenzie says: "The fact is, that croup is a disease which commonly commences in the pharynx, and only in about 10 or 12 per cent. of cases originates in the larynx or trachea." In a large number of cases of pseudo-membranous croup, I observed little islands of exudation of a pearly lustre in the neighborhood of the glottis, but that appeared at the same time as the exudation in the larynx, and sometimes even later. Very seldom have I observed the pharyngeal exudation preceding the laryngeal, certainly not more than in five per cent. of the whole number. No one could confound the water-white of the croup exudation with the milk-white of the diphtheritic one. Dr. Mackenzie adds, "Difference of site, moreover, in a constitutional disease, does not constitute a specific difference. Here the constitutional nature of pseudo-membranous croup is *assumed* to be *the same in kind* as the constitutional nature of diphtheritic croup—to me they seem to differ as much as croupous pneumonia differs from gangrene of the lungs.

"My idea of the problem to be solved is, in fact, this: It must be admitted that the diphtheritic poison is capable of giving rise to a plastic inflammation of the larynx, apart from the existence of any similar affection of the pharynx. But there is good reason to believe that during epidemics of diphtheria, the cases in which this occurs are, in the highest degree, exceptional. If, therefore, it can be shown that in the practice of a general hospital the cases of plastic laryngitis, of uncertain origin, bear a large proportion to the total number of cases of diphtheria, there will be a strong probability that the majority of the former cases are dependent upon some other cause than the diphtheritic poison"—(*Diphtheria and Croup*, by W. H. Lamb, M. B., and C. Hilton·Fagge, M. D., "Guy's Hospital Reports," 1877).

True croup is most frequently caused by the sudden passage from warm to cold air, and is often occasioned by sleeping in very cold bed-chambers after having been all day in hot rooms. Diphtheritic croup is the manifestation in the larynx of a blood disease ; and whatever effect external cause may have in bringing about diphtheria in general, they have very little in producing the laryngeal variety. True croup almost invariably begins with catarrh and fever, and this in exact ratio with the severity of the local symptoms; difficulty in swallowing is very rare, is always very slight when it does occur, and it is always dependent on the laryngeal affection. In diphtheria catarrh is rare, for the fetid sanies which flows from the nostrils and mouth can hardly be called catarrhal, and from the very inception signs of deep-seated constitutional mischief are evident ; sore throat and difficulty in swallowing precede the laryngeal affection.

Dr. Edmonds, of St. Louis, points out that while "croup is bold, abrupt and, as it were, outspoken in manner and character, diphtheria is sneaking, insidious and undefined in its mode of approach," and this, so far as my experience extends, holds almost universally good. Dr. L. Fleischmann remarks that while croup most frequently affects the mucous membrane of the air passages, diphtheria frequently has "multilocular invasion," the fauces, nose, genitals, intestines and the skin being affected simultaneously. Again, the kidneys and intestines are normal in croup, while in the laryngeal form of diphtheria they are often involved.

Dr. Hillier remarks, " It appears to me as impossible to maintain that croup is merely a local disease, as that pneumonia is merely local, or catarrh, both of which are generally indications of a morbid constitutional state." Precisely so, croup and pneumonia are strictly analogous diseases, and the "morbid constitutional state" is, in both affections, dependent on the local disease. But that may be freely conceded without granting the identity of true croup and diphtheritic croup.

Dr. Morell Mackenzie says, "It is true that in croup the general symptoms are not so severe as when the membrane is thrown out on an extensive portion of the pharynx. This fact admits of ready explanation, on the view that the septic symptoms are in part secondary to the local processes." But in diphtheritic croup, the septic symptoms precede the local process in the vast majority of cases, and I have rarely seen diphtheria attack the larynx in the first instance. Again Dr. Mackenzie says, "When the primary septic poisoning is powerful, the constitutional symptoms are, however, as marked in so-called croup as in diphtheria." No other medical observer on either continent, so far as I know, has asserted that pseudo-membranous croup is accompanied or followed by constitutional symptoms similar to those which accompany diphtheritic croup, for, as Dr. Edmonds ably points out, "In diphtheria the diseased appearance in the larynx or trachea is simply the outcropping of a previous constitutional taint. The outcropping is not confined to the larynx or trachea, but may show itself in the nose, throat, eyes, ears, and even upon the cutaneous surface, wherever there may be the slightest break of integrity or denuding of surface."

True croup, then, is a local inflammation of an exudative character accompanied by a purely inflammatory fever, while diphtheritic croup is originally a blood poisoning, and the laryngeal disease is what Hahnemann would call "a local manifestation" of that blood poisoning—hence, in a majority of cases of true croup the pharynx is healthy, or almost healthy, while in a majority of cases of diphtheritic croup the pharynx is diseased.

Another important distinction between the two diseases

is, that while true croup is a sthenic inflammation, diphtheritic croup is accompanied by fever of an adynamic type. In opposition to this, Dr. Morell Mackenzie asserts that "cases of sthenic croup are very rarely met with, and the same remark applies to diphtheria." All the text-books, all the observers of both continents, describe the fever of pseudo-membranous croup as being sthenic, and I venture to say that Dr. Mackenzie stands alone in his position, and though he says that "distinctions based upon differences of type in the two diseases can have no weight," it is quite certain that all diagnosis depends upon the detection of just such "differences of type." While, then, true croup commences with inflammatory fever of a very pronounced type, diphtheritic croup is accompanied by fever of a typhoid or adynamic type.

Another important difference is that in croup the cervical lymphatic glands are not swollen, while in diphtheritic croup the cervical glands are inflamed and consequently enlarged. Dr. Mackenzie admits that "the cervical glands are not often affected in croup, because the mucous 'membrane of the larynx has no communication with the superficial cervical glands; on the other hand, as stated above, there is an elaborate connection between the pharynx and the lymphatic glands," and he quotes Luschka's ingenious explanation: "Whilst the lymphatics of the mucous membrane of the soft palate, of the tonsils and of the back of the pharynx have very free communications with the numerous glands below the angle of the jaw, the absorbent vessels of the mucous membrane of the larynx and trachea are conveyed only to the solitary gland just below the greater horn of the hyoid bone, and the small gland at the side of the trachea." The fact is, that in pseudo-membranous croup the cervical glands are swollen, especially those at the outer border of the sterno-mastoid muscle; but, as Dr. L. Fleischmann accurately points out, in croup there is swelling of the glands, but almost never suppuration of fetid character, while in diphtheria suppuration of the glands is of frequent occurrence.

Another point of difference, formerly looked upon as all but conclusive, and still much relied upon by excellent observers, is that while albuminuria is present in diphtheritic croup, it is *not* present in pseudo-membranous croup. Even as late as 1880, Dr. Henry Hartshorne, of Philadelphia, affirms that "croup is not followed by albuminaria," and Dr. Squire states that in drawing the diagnostic lines between the two morbid states, "*the presence of albumen in the urine is conclusive.*" On the other hand, Dr. Morell Mackenzie affirms that "in croup albuminuria is often found." Dr. Hillier says that "albumen has been found in the urine of patients with croup;" while Dr. Alfred Meadows, of London, states that "in mild cases of diphtheria there is often no albuminuria, while in severe cases of croup it is not unfrequently present." But the force of all these observations, undoubtedly correct as they certainly are, is broken by the fact that the elaborate researches carried on under the auspices of the Royal Medico-Chirurgical Society, of London, conclusively prove that albuminuria occurs in cases of laryngitis in which no membrane is formed—that is, in simple laryngitis which is neither croupous nor diphtheritic. The conclusion seems to be that albuminuria is relatively less frequent in pseudo-membranous croup than in diphtheritic croup, but that as it also occurs in phases of laryngeal disease which belong to neither of these classes, this criterion must be looked upon as being indecisive.

One of the principal clinical differences is that while paralysis occurs after diphtheritic croup, it does not occur in true croup. Dr. Squire says that "paralysis of some of the muscles of vocalization, deglutition or of motion, is equally distinctive of diphtheria;" Dr. L. Fleischmann states that in croup "paralysis never occurs," while in diphtheria, "even in mild cases, severe nervous disturbances are common;" and J. Solis Cohen gives it as a chief point of difference: "In croup no secondary paralysis; in diphtheria secondary paralysis is frequent."

On the other hand, Dr. Morell Mackenzie says that

"paralysis is rare in croup, because nearly all the cases terminate fatally, but it is occasionally met with in those that survive;" I have never met with paralysis after true croup, and no writer on the disease has ever mentioned such a thing, and Dr. Mackenzie is describing cases which nineteen medical men out of twenty would call diphtheritic croup. Dr. Hillier remarks: "Even when epidemics of diphtheria prevailed in former times, the nervous sequelæ were not noted; we have no record of these phenomena till a comparatively recent period. It is quite probable that even if symptoms of disordered innervation had followed sporadic croup in as large a proportion of cases as they follow epidemic diphtheria, they would not have been connected with the previous illness," but some of the Spanish and Sicilian writers describe nervous phenomena following *garrotilla*, and it would be strange, indeed, if observers failed to connect disordered innervation with true croup in the purely hypothetical case stated by Dr. Hillier. Dr. Alfred Meadows observes: "Paralysis is frequently absent, even in rather severe cases of diphtheria. In France it is said that paralysis is present in at least one-third of the cases, while in England it does not occur more frequently than in ten per cent. of the cases; therefore, it might be argued that at least in those cases where this symptom is absent, there is nothing essentially distinguishing it from croup." Dr. Meadows does not affirm that paralysis has ever been observed to follow true croup, and in cases of diphtheritic croup, when paralysis is absent, other diagnostic marks, equally conclusive, are present. Another writer, who upholds the doctrine of the unity of these diseases, candidly admits that "sporadic cases"—by which he means cases of true croup—are rarely followed by paralysis or albuminuria, and this is the conclusion of a vast number of careful and impartial observers.

True croup is non-epidemic, non-contagious and non-inoculable, while diphtheritic croup—in common with all manifestations of diphtheria—is epidemic, contagious and inoculable.

In 1879 the Royal Medico-Chirurgical Society of London appointed a committee to examine the relations existing between croup and diphtheria, and in the very interesting report the following passages bear on this question: "Membranous inflammation, chiefly of the larynx and trachea, to which the name 'membranous croup' would commonly be applied, may be imparted by an influence, epidemic or of other sort, which in other persons has produced pharyngeal diphtheria. And, conversely, a person suffering with the membranous affection, chiefly of the air-passages, such as would commonly be termed membranous croup, may communicate to another a membranous condition, limited to the pharynx and tonsils, which will be commonly regarded as diphtheritic." Gerhardt asserts that he has, himself, " described a sporadic case which proved contagious," which may well be the case if the sporadic case was diphtheritic. Dr. Meadows observes, " we know that croup does sometimes occur epidemically, and if it be not contagious so neither is it certain that diphtheria, when it attacks a number of persons in the same house or locality, is really communicated; for the fact may be due to the exposure of those affected to the same influence and at the same time." To this it is sufficient to reply that it is quite certain that diphtheria is contagious, but no sufficient evidence has yet been adduced to prove that croup possesses the same quality, and the *onus probandi* lies with those who assert that it is contagious.

Dr. Edmonds says, " Diphtheria is a zymotic, constitutional blood disease, is in many instances believed to be contagious, is undoubtedly inoculable by application of matter from a diphtheritic part to the mucous or denuded surface of a healthy subject." Dr. Charles West points out that while " croup is influenced by climate and season, is endemic in some localities, but not epidemic nor contagious, diphtheria is independent of climate or season, contagious and often epidemic." Dr. L. Fleishmann observes that while " in croup there is no infection of the blood, with corresponding symptoms depending thereon, in diphtheria there is infection of the blood and fatty degeneration of the striped, muscular tissue, especially that of the heart, and he adds that while croup is not inoculable, diphtheria *is* inoculable. Hirschel's experience is that while diphtheria is contagious and mostly epidemic, croup is not contagious, and is mostly sporadic.

Steiner's views on this point are valuable, and he concludes that "the attempt to distinguish croup and diphtheria as two entirely distinct diseases has been unsuccessful, both from an anatomical and from a clinical standpoint," yet he admits that "primary true croup is *not a contagious disease*, although it is so regarded by Bohn, Gerhardt and others. Diphtheritic croup, however, possesses this quality in a high degree." Dr. Hillier, too, asserts that "*epidemic croup*" *is always diphtheria*—which is precisely the conclusion of the present writer.

True croup is a disease of cold weather; diphtheritic croup ravages in all weathers and in all seasons. Croup is caused almost solely by climatic influences, although at times it is endemic; diphtheria is favored by everything which promotes the growth of spores, as crowded dwellings, personal uncleanliness, and so forth.

True croup rarely occurs more than once in the same patient, but it is quite common for children to have several attacks of diphtheria. As to age, true croup appears much earlier than diphtheritic croup. True croup is almost peculiar to children; adults as well as children are the victims of diphtheritic croup. Dr. Alfred Meadows says: "In regard to croup seldom or never attacking adults, while diphtheria frequently does, this can hardly be relied upon, because the same may almost be said of scarlatina." Adults are seldom attacked with scarlatina, simply because they have passed through that ordeal in youth; would Dr. Meadows affirm that adults are free from croup *because* they have had it in youth? Dr. Hillier ingeniously argues that "when diseases become epidemic they are more liable to attack adults, who escape when the disease is only sporadic," but this is a theory as yet unsupported by facts and figures.

I have never observed in true croup the scarlatiniform eruption so often seen in diphtheritic croup, and Dr. Lyon, of Connecticut, points out that in croup pseudo-membranes of the skin are never observed, while in diphtheria pseudo-membranes of the skin are occasionally observed.

Dr. Ludlam, of Chicago, writes as follows: "The dyspnœa in croup is paroxysmal, and invariably worse at night. There is a true spasm of the laryngeal and tracheal muscular fibres. At intervals the patient breathes almost naturally. In a few moments, especially if permitted to sleep, he is in a fit of suffocation again, which, by-and-by, alternates with relative repose. The ease and freedom of the respiratory movements in diphtheria vary considerably at intervals, but the intervals occur irregularly during the day as well as at night, and the relief afforded by them is less marked than in the case of croup. In true croup a trembling, vibratory sound may often be heard on auscultation, denoting the presence of floating false membrane, but the characteristic auscultatory indication of diphtheritic croup is a soft gurgling sound, similar to the cavernous râle of phthisis. Dr. J. Solis Cohen remarks that in croup there is no weakening of the heart's action; the pulse frequently strong and hard, while in diphtheria there is marked weakening of the heart's action; pulse never strong and hard, even though rapid and full.

Finally, it is a matter of frequent observation that while in croup the general health is rapidly reëstablished, a complete recovery, without sequelæ, following the cure of the local disease, in diphtheritic croup the convalescence is remarkably slow and tedious, with annoying sequelæ lasting for months and even years.

Monti considers croupous laryngitis to be a separate disease, independent of diphtheria, but he considered that it *may* arise from diphtheria, and Dr. J. Solis Cohen admits that there is no actual anatomical distinction between croup and diphtheria, though he contends that the clinical differences are numerous and important. Dr. Charles West, a most distinguished writer, has come to the conclusion, which he long hesitated to adopt, "that what differences soever exist between croup and diphtheria, they must be sought elsewhere than in the pathological changes observable in the respiratory organs." Nevertheless, he adds, "If we extend our inquiry beyond the mere changes wrought in the respiratory organs, the differences between croup and diphtheria at once become apparent; and the affinities of the latter disease are seen to be to the class of blood diseases, rather than to that of purely local inflammations to which croup belongs."

The Committee of the Royal Medico-Chirurgical Society, of London, came to the conclusion that "these two diseases are identical," but, strange to say, the discussion on the subject subject did not lead the members to harmonious conclusions, for not very many English practitioners would say with Dr. Hillier, "I can detect no distinction between membranous croup and laryngeal diphtheria." Pseudo-membranous croup and diphtheritic croup are not identical diseases, but they certainly have much in common, and, as Dr. West puts it, "the sameness of the anatomical changes produced by two diseases does not suffice to establish their identity." He adds, " The practitioner of midwifery knows that simple puerperal metritis and puerperal fever are diseases which differ widely in their symptoms, their course, their danger, and the degree in which they are amenable to remedies, though in both, when they terminate fatally, precisely the same alterations in the womb are discovered."

It is impossible to confound typical cases of true croup with typical cases of diphtheritic croup, and only occasionally need there be any doubt as to the diagnosis. At the same time, I concede that since the advent of diphtheria in Canada (1858) I have observed an increasing disposition in pseudo-membranous croup to take on *a kind of diphtheritic aspect*, but the same thing has been noticed with all diseases of the mouth and fauces, and one of the very strongest proofs of the essential dissimilarity of the two diseases is that given by Dr. Henry Hartshorne: "A table is given in Meigs and Pepper's treatise on the Diseases of Children, which shows that after diphtheria had, about 1860, become recognized as, at that time, a new disease in Philadelphia, the mortality from it added, for several successive years, more than 300 to the deaths in each year in that city, while the deaths from croup continued to number annually, as before, from 200 to over 400."

The prognosis is very bad. The *Lancet Sanitary Commission-Report* on diphtheria states that "symptoms of croupal suffocation soon supervene from the extension of the diphtheritic formation to the air passages, and when this is the case, recovery is exceptional." Dr. Churchill says that "when the false membranes extend into the larynx and trachea we shall have croup with all its danger." Condie remarks that "when the disease extends to the larynx, it is very frequently fatal." According to Greenhow, "comparatively few persons recover when diphtheria extends downwards into the air passages; but sometimes moulds of the larynx, trachea and bronchial tubes, to their third or fourth division; and in a case seen by Mr. Thompson, of Launceston, to the fifth division, are expectorated with immediate, though too often only temporary, relief to the patient, who frequently succumbs from a renewal of the exudation."

Dr. J. F. Meigs says, "If it extend into the air passages very soon after the invasion, it may cause death within a few days." Sir George Duncan Gibb thinks that "the prognosis of this form (laryngeal diphtheria) is extremely unfavorable." Dr. Bernhard Bæhr says that "the extension of the diphtheritic process to the larynx and lungs is almost always fatal." Maunsell affirms that "in very acute cases the false membrane will spread into the larynx, if not early arrested; and in some instances its formation seems to occur almost simultaneously in the air passages and on the pharynx, the croupy symptoms appearing to co-exist with the appearance of lymph on the fauces; such an event we need hardly say must be almost necessarily fatal."

Vogel states that "in diphtheritic croup, especially after measles, a recovery now and then takes place, upon which the treatment, as we will see further on, has no very remarkable influence. Where collapse, cyanosis and an uncountable pulse have supervened, there speedy death may be prognosticated with certainty." Oehme says that diphtheria of the larynx has proved, in the greater number of cases, a fatal disease. Some physicians have not hesitated to say they have never cured a case." Jacobi says that "diphtheria of the larynx, whether it be of primary origin or the result of extension from the fauces, is *nearly always fatal*. In severe epidemics the mortality is 95 per cent.

In the year 1860, I wrote an essay on diphtheritic croup in which I expressed myself as follows in regard to the prognosis: "But little need be said as to the prognosis of this disease; *it is bad, very bad*, and I do not believe that more than one-half of the cases recover, even under the best homœopathic treatment. The cause of this is evident, for before the larynx is attacked, the patient has usually been depressed and worn out by the primary disease, and is quite unfit to contend with such a formidable foe. I find that the best plan is to explain the state of matters frankly to the parents on being called to the case, and here, as in many other circumstances of life, 'honesty is the best policy.'" The younger the child the greater the danger of diphtheritic croup coming on in the course of diphtheria. Again, the younger the child the greater the danger when laryngeal diphtheria does make its ominous appearance, and this increased danger arises from the small size of the larynx in infancy. Romberg states that "Richerand was the first to determine that the larynx and glottis, which in early life are very small, suddenly increase at the period of puberty, in the male sex in the proportion of 5.10, in females of 5.7. Schlemm has confirmed this observation, and has added a few details: thus he found the rima glottidis of a child of twelve years one and a half to two lines longer than that of a child of three years, and in the latter it was three-quarters of a line longer than in a child of nine months." When the larynx, trachea and bronchial tubes are lined with diphtheritic membrane the case is all but hopeless, though strange cures take place, even when the patient seems to be *in extremis*, and I particularly remember one notable case in which, under the influence of Kali bichromicum, the patient expectorated a cast of the larynx, trachea and larger bronchial tubes with immediate and permanent relief.

In the first volume of Marcy and Hunt's Practice, page 763, we read that "a persistent use of the proper homœopathic remedies will cure nearly all cases of this malady" (diphtheria), and to this somewhat startling statement they add that "we have treated more than 200 cases, including many of the malignant type, and our losses have not been one per cent." It is not stated how many of these two hundred cases were laryngeal in their nature, though one would like very much to know, but I imagine that no experienced physician would make these statements in connection with diphtheritic croup.

Opinions differ very much as to the value of tracheotomy in diphtheritic croup, and till quite recently Dr. Slade was almost the only writer of eminence who spoke favorably of it. He says: "Without going into a history of tracheotomy, or a recapitulation of the arguments on the one side or the other, we most unhesitatingly say that, under the circumstances above mentioned, this operation is a resource which we are in duty bound to employ for the safety of our patients, and in view of what experience teaches us is otherwise certain death. It is not by so doing that we increase his chances for life solely, but in case of an unfavorable termination we render his last moments less distressing." Dr. Squire, who sharply distinguishes pseudo-membranous croup from diphtheritic croup, writes as follows: "Tracheotomy should be performed whenever the increasing recession of the softer parts shows that the cause of obstruction to the entrance of air is increasing. In the greater number of cases, if the local indication of the glottis and larynx do not suffice to obviate the danger, tracheotomy, performed early, is much more likely to be successful than after the use of remedies that in any way impair the vital powers. A delay that admits of secretions accumulating in the bronchi is dangerous, and extension of the disease to the lung is the one insurmountable obstacle to success. Where the effects of the obstruction are more suddenly induced, tracheotomy, performed at the very last moment of apparent life, may save it. No degree of severity in the general disease should interfere with this means of arresting threatened death from asphyxia, unless the presence of some other complication, necessarily fatal, can be demonstrated. I recently saw a case in consultation with Mr. Adams, in which, had it occurred at the commencement of the epidemic instead of towards the end, I should have decided against tracheotomy, concluding that it must end fatally, although unconsciousness had set in before commencing to operate; the child, six years old, recovered."

Professor Rosen, of Tubingen, reports forty-two cases of tracheotomy in diphtheritic croup, with nineteen recoveries. In six of the cases asphyxia had advanced too far before the operation, and of the subsequent deaths, one took place from pneumonia, fifteen days after, and one from albuminuria in the third week. Professor George Buchanan, of Glasgow, asserts that in every eight cases of tracheotomy performed on children practically moribund from suffocative membranous effusion into the trachea, *he has saved three.* The Professor reports 50 cases of tracheotomy in the *British Medical Journal* (1880), of which 17 were classed as croup and 33 as diphtheria, the latter including all those forms in which there was a distinct deposit of white false membrane on the tonsils, palate, or fauces. Of the 17 croup patients, 10 died, 1 immediately after the operation, the others in from 3 hours to 4 days. Of the 33 diphtheria patients there was a mortality of 21, 1 of whom also died immediately after the operation, the others in from 6 hours to 13 days.

Dr. W. H. Day, who, like Dr. Squire, holds that the two

diseases are totally distinct, yet reports the following deeply interesting cases: "Two interesting cases of successful tracheotomy, in the last stage of diphtheria, were brought before the Clinical Society by Mr. George Lawson and Mr. Pugin Thornton (Feb. 28, 1879). Two cases of diphtheritic laryngitis have been recorded, in which recovery also followed tracheotomy. The first case was that of a boy six years of age, who was admitted into the Middlesex Hospital under the care of Dr. Coupland, May 30, 1880. The successful issue was owing to the operation having been performed at an early period of the disease before much false membrane had formed.

"The second case was also that of a boy seven years of age, who was admitted into the Children's Hospital under the care of Dr. Gee, on September 15, 1879. Recovery followed quickly, notwithstanding the extreme dyspnœa at the time of operation, and the large quantity of membranous casts expelled through the tube afterwards."

Steiner thinks that "when the larynx becomes implicated the various external and internal remedies, which have already been referred to under the diseases of the larynx, such as emetics, must be employed, and, failing any benefit from these, there remains only the operation of tracheotomy." Dr. Jacobi, the most persistent advocate of the unity of croup and diphtheria on this continent, writes as follows: "In regard to tracheotomy, that last resort in croup, I cannot refrain from stating that, in proportion to the increasing severity of the diphtheritic epidemics, the results of tracheotomy in my hands and in those of others, have grown worse and worse. Of sixty-seven tracheotomies which I published twelve years ago, twenty per cent. recovered; about two hundred tracheotomies performed by me since that time brought down the percentage of recoveries to such a low figure that only the utter impossibility of witnessing a child's dying from asphyxia has goaded me on to the performance of tracheotomy. I here add that I do not wish it to be inferred that I have changed my views concerning the indications for the operation of tracheotomy, as Boehme seems to believe. On the contrary, in spite of numerous ill successes, I hold to the principle that where there is danger of suffocation through stenosis of the larynx, there is the indication for tracheotomy. Where there is no stenosis, I am glad not to operate. The results are not so bad, after all, when we remember that only such cases are operated upon which would be sure to die, if the operation were not performed."

One would like to know how many of the patients treated by Buchanan, Squire and others had true diphtheria with blood-poisoning, and how many had a purely local laryngitis with exudation, but lacking the blood-poisoning which is almost of the essence of diphtheria. On this point the remarks of Dr. Alfred Meadows are of the greatest value, the more so as that excellent writer believes croup and diphtheria to be identical diseases. After stating that Trousseau believes that half the operations performed will be successful, always provided that tracheotomy takes place when the chances of cure are possible, he adds: "This restriction is important; for if the diphtheritic infection is thoroughly rooted in the system, if the skin, and particularly the cavities of the nose, are invaded by this special phlegmasia, as is often the case in France; if the quickness of the pulse, delirium and prostration indicate a profound poison; and if the danger is rather in the general state than in the local lesion of the larynx or trachea, certainly the operation should not be tried, for it is invariably fatal." Dr. Meadows further adds, "when, however, the local lesion constitutes the principal danger of the disease, no matter to what degree asphyxia has arrived, even when the child seems to have only a few moments to live, tracheotomy very often succeeds"—*and this " local lesion" is correctly styled pseudo-membranous croup.*

Dr. Morrell Mackenzie thinks that "considering the enormous mortality of laryngeal diphtheria, even the most unfavorable figures prove that in such cases tracheotomy is not only justifiable, but that it is a positive duty;" yet Mackenzie gives a table of "operations for croup," in the *Hôpital des Enfants Malades*, showing that whereas in 1851 the cures were 1 in 2.21, in 1875 the cures were 1 in 4.76. Mackenzie gives a similar table for the *Hôpital Sainte Eugenie*, showing that while the cures in 1854 were 1 in 4.50, in 1876 the proportion of cures was only 1 in 8.31. M. Mazard attributes this steady increase in the mortality, after tracheotomy, " partly to the progressive extension of the operation to more and more hopeless cases, and partly to the more malignant character of the disease in Paris during recent years;" but I incline to attribute it to the fact that in the earlier years true croup was chiefly present, while in the later years there was much more diphtheria.

Oertel writes : " According to the notes of Professor von Nussbaum, which he has most kindly communicated to me, of twelve undoubtedly diphtheritic children, whose ages varied between three and four, and on whom he had performed tracheotomy, all died ; and only two older ones, whose ages were twelve and fourteen, survived, but in them the whole course of the disease had shown itself much more favorable." Dr. Helmuth remarks : " The results after the operation of tracheotomy in croup are not very satisfactory, and in diphtheria they are, as far as my observation, reading and experience extend, still less so. I can call to mind but four cases in which the operation has been performed in diphtheria, one of which is said to have proved successful. None of these cases, however, occurring in my own practice, and circumstances occurring which have prevented actual inquiry in reference to the minutiæ of each, I am not prepared to offer any remarks upon them. But in the cases which I have treated, and which have succumbed to the diphtheritic poison, I have not witnessed one which would justify the interference of the surgeon." Dr. A. L. Voss of New York says : " It is worthy of remark that I have not heard of a successful operation in New York during the year 1859, famous for diphtheria. Tracheotomy is a *dernier resort* in diphtheria. I have no confidence in it in this disease. Diphtheria is an affection of which the local lesion is the least important part. Its erratic nature, its proneness to reappear upon a neighboring or remote surface, argues very strongly against the promise of success by local means alone. If you remove the plastic deposit from the trachea by a surgical operation, a few hours later will be apt to reveal symptoms of a like formation within the larynx or the bronchi. Possibly the operation may be serviceable in croup, but not in diphtheria."

In almost all the statistics adduced to prove the value of tracheotomy in diphtheritic croup, the comparatively manageable pseudo-membranous croup is mingled with the very unmanageable diphtheritic croup, so that little reliance can be placed on them. For example, Dr. Morell Mackenzie writes: "At the Hospital for Sick Children in the twelve years, 1864 to 1876, sixty cases of croup and diphtheria were operated on. Of these, thirteen, or 21.6 per cent., were successful." One wants to know how many of these cases were croup, and how many were diphtheria, for the mortality is much more than trebled by the presence of the diphtheritic blood-poisoning, and no experienced physician expects to save 21 out of every 100 cases of genuine diphtheritic croup. And even when some little attempt is made to discriminate between the two diseases, the diagnosis is so superficial that it commands no respect. Thus of the 50 tracheotomy cases of Buchanan's, already alluded to, 17 were classed as croup and 33 as diphtheria," the latter including all those forms in which there was a distinct deposit of white false membrane on the tonsils, palate and fauces." Of the 17 croup cases 10 died, of the 33 so-called diphtheritic ones 21 died—not a very great difference in the mortality. And it is not sufficient for diagnostic purposes to tell us that there was a "distinct deposit of white false membrane on the tonsils, palate or fauces," for that is often seen in pseudo-membranous croup, but one wants to know the history of these cases, whether or not diphtheritic blood-poisoning was present, and till that is done, I will ponder the words of Vogel, "*We really have few diphtheritic, but mostly genuine fibrinous croup patients.*"

Commenting on von Nussbaum's cases, already alluded to, Oertel writes as follows, and his weighty words must command the respectful attention of all physicians of experience: "If now, figures are to be found in literature which furnish much more favorable statistics of tracheotomy in diphtheria, these data cannot be considered as trustworthy so long as the boundaries between croup and diphtheria are not precisely defined; in the cases referred to above the diagnosis of diphtheria was established beyond a doubt. It is very evident that the issue of such an operation should be wholly different if the case is one of a simple exudative process in the respiratory mucous membrane, following a local inflammation of high degree, and not one in which the local trouble is the primary localization of a general infectious disease."

In diphtheritic croup, as a very general thing, the condition of the patient forbids any surgical interference whatever, and it is only in rare cases like those of Squire and von Nussbaum that it should be even thought of. In croup you have to do with a disease which is local, or at least very largely so, and here tracheotomy is admissible in certain cases, but in diphtheritic croup you have to do with a patient who, in addition to a severe local disease, is suffering from a violent blood-poisoning, and here the chances of success, or even of palliation are, save in very rare cases, illusory in the extreme.

The bed-chamber must be lofty and well-ventilated. The air must be both warm and moist, and draughts must be carefully avoided. The only prophylactic treatment which is of any avail is the prompt removal of children from the infected locality. That must be done at once, for here no chemical agent is of any use, and some of them, notably chloride of lime, positively invite the disease to the larynx. The food must be nourishing and liquid. Milk very slightly thickened with arrow-root is excellent, and well made beef-tea is always in place. But I have had the best results from oyster-soup, giving only the thin part. Both English and German writers give stimulants in large doses, but I have rarely seen any good from them, and it seems to me that the dose of which Oertel speaks—an ounce or an ounce and a half of Cognac in twenty-four hours to children only three or four years old—is excessively large. In the matter of after-treatment, it is of importance to see that the child does not over-exert the organ of voice. Rest, nourishing food, fresh air are indispensable, and, when seasonable, sea-air is the best of all restoratives.

Physicians of our school differ very much as to the value of Kali Bichromicum in laryngeal diphtheria. Dr. Ludlam thinks that it is almost specific to the diphtheritic membranes found upon the respiratory epithelial surface, and this is endorsed by Drs. Marcy and Hunt, while Dr. Lord, of Chicago, emphatically says that it is *the remedy*. Dr. A. E. Small, of Chicago, says : " I have found the 3d attenuation of this remedy of the greatest value in diphtheritic croup when administered early after the manifestation of the difficulty. I have given it when the following symptoms were present : hoarse, croupy cough ; sore throat ; the appearance of livid patches, indicative of false membrane, at the posterior of the fauces ; great prostration and laborious breathing." On the other hand, Dr. Hughes, of London, has used it " without the least benefit," and in a later work, he adds that "in laryngeal diphtheria it does all that medicine can do, which, unhappily, is not much," and Dr. Laurie seems to consider it a kind of forlorn hope to be given if Bromine should fail ; the Bromine have been given when Iodine failed. Dr. Bernhard Bæhr, while admitting that " the symptoms of this drug undoubtedly point to its use in diphtheria, and assign to it an important rank among the remedies for this disease," remarks that " striking therapeutic results have not yet been attained with it." In this matter I entirely and cordially agree with Dr. Ludlam, and have no hesitation in assigning to Bichromate of potash the first place in the brigade of remedies with which we combat diphtheritic croup. Dr. Ludlam's remarks are worthy of the most careful perusal:

"1. This remedy seems especially appropriate to pseudo-membranous lesions of a diphtheritic nature affecting the respiratory mucous surfaces, as the nares, the superior portion of the pharynx, the larynx, the trachea, and the bronchial tubes, even down to their ultimate ramifications.

"2. Where the deposit is of firmer texture, more apt to be developed into casts which are cartilaginous or pearly in in appearance, elastic, fibrinous, and more securely attached to the subjacent integument.

"3. It is indicated in all those cases where a transfer of the local disorder to the larynx or trachea impends, as shown by soreness of the larynx when pressed upon from before backwards, aphonia, croupy inspiration or cough, and a desire on the part of the patient to lie with the head thrown backwards in order to open the glottis.

"4. It may also be given with excellent results in case the tonsils are almost or quite enveloped by a thick and well-organized deposit, and in which at the same time the patient has an almost incessant cough.

"5. Also where, with the foregoing symptoms, there is an evident tendency to ulceration and deposit upon remote mucous membranes, as, for example, those of the uterine system. In my own experience, the Bichromate is almost a specific to diphtheritic formations upon the free uterine, and to those found upon the respiratory epithelial surfaces.

"6. Since in all these cases the putrid symptoms are less marked than in the pharyngeal and alimentary diphtheria, you should take the hint to cease the employment of Bichromate when these symptoms ensue. The Iodine of arsenic, Nitric acid, or Carbo-vegetabilis are much more decidedly indicated for the relief of such a condition."

The diphtheritic croup in which this remedy is indicated, is generally an extension from the fauces, simply because that is the usual development of the disease, but it is the first remedy to be thought of when the disease originates in the larynx in the first place, and extends thence to the fauces. The pseudo-membrane lining the fauces and extending to the air passages is whitish-yellow or of an ashy-grey hue, and the fetor is quite marked. The croupous cough occurs in paroxysms, especially worse from two to three o'clock in the morning, and this cough occasionally expels viscid mucus, which may be drawn out into long strings, and the same tough and stringy discharge occasionally appears in the nostrils, forming masses of partially-dried mucus, or there may be a thick, dark, bloody discharge from the nostrils. The tongue is raw, red and shining, or covered with a brownish-yellow coating, and the parotid and submaxillary glands are distinctly swollen. Upon deglutition, the pain shoots up the ear and to the neck of the affected side. The patient is very weak, and has a cachectic look.

Dr. Lord, of Chicago, advises the administration of this remedy by inhalation. The following is an extract from his report of a very severe case in which the remedy caused an aggravation, even when given in moderate doses: "When the Bichromate was given at intervals of an hour or more, the patient uniformly got worse. The cough was almost constant, except in the night when asleep. It ran up from a slight hacking to suffocation, which was only prevented by a means which I have purposely omitted to mention that I might direct your attention more particularly to it. After the twentieth day, when the cough became dry, and the respiration whistling, and when suffocation seemed imminent, inhalations of the Bichromate were used with prompt relief; of course it was only temporary, but it was a respite. But for it death must have ensued. It did not fail in a single instance of easing the breathing and loosening the cough, and ejection of membrane or large quantities of stringy mucus followed. The method was simple. Two or three grains of Bichr. 2 were put into a small tin teapot and half a teacup of hot water poured on. The vapor passing from the spout was inhaled. I do not think that any medicines given in this case but the Arsenic and Bichromate had any good effect. I was so well satisfied of this that in all subsequent cases I have trusted entirely to the Bichromate as the specific remedy, and have had no reason to repent it. Other remedies may be required, but that is the remedy."—(*Illinois State Hom. Trans.*, 1862.)

Dr. Hughes says: "I think that Dr. Neidhardt's suggestion is very good, that it is necessary to attack the poison in the blood, even while, by the medicines specifically affecting the air passages, you are combating its dangerous local manifestation. He usually administers the Bichromate of potash (1st trit.) in alternation with his Chloride of lime. He has recorded two instances in which this treatment proved successful." *To me, cases in which alternated remedies were used are as if they never had been recorded.* This remedy seems to act best in the 2d or 3d decimal triturations, a small powder in half a cup of water, of which a teaspoonful is to be given every hour or oftener. The remedy should be given by inhalation at the same time.

All who have used Iodine consider that it is best indicated in the early stage of diphtheria when much glandular irritation is present, and when the disease threatens to attack the larynx. Dr. Laurie thinks that "when in addition to the formation of specks or patches of exudation of greater or less extent, with sore throat, enlargement of the tonsils or glands of the neck, disinclination for food, *difficulty of breathing, cough and alteration of the voice ensue,* the administration of Iodine should be at once resorted to." Dr. Kidd says that the essential pathogenetic action of Iodine comes the nearest of all our remedies to the essential characteristics of diphtheria in its constitutional and local manifestations. Dr. F. G. Snelling says that its internal use should be in frequent repetition, and accordingly he advises that ten drops of the first decimal dilution of Iodine be added to half a cup of pure cold water, a teaspoonful to be given every 20 or 30 minutes. "To produce a prompt and perfect influence, Dr. Kidd thinks it best to administer it, '*similia similibus curantur*,' in the mode of entrance of the disease itself—viz., by inhalation; or the Iodine, in substance or in tincture, may be placed in an open vessel near the patient, as it is thus slowly evaporated, and mixes with the air in a highly divided and quickly acting form."—(*Snelling.*) Dr. McNeil, the best writer on the homœopathic therapeutics of diphtheria that has yet appeared, says that,, it is only in rare cases that Iodine will ever be indicated "—and I entirely agree with him.

Dr. Helmuth supplies us with the following valuable details as to the best mode of administration by inhalation : "When the disease is not arrested by these medicines. (Caustic ammonia and Protiodide of mercury)—and there is the slightest appearance of cough—I order the inhalation of the *vapor of Iodine*, and that medicine in the second dilution, in water, every two hours. The inhalation is conducted as follows: a small teapot is filled with boiling water, and a teaspoonful of pure *tincture of Iodine* poured therein; the patient takes the spout of the vessel in the mouth, and the head being covered with a towel, a few inspirations are made. This method is resorted to when there is no inhaling glass convenient, and it will be found to answer the purpose exceedingly well. The inhalation may be repeated three times during the day. There can be no doubt of the efficacy of this method of treatment—viz.: *Iodine*, internally and topically, by inhalation, in severe diphtheritis, even after the cough has *commenced*. I have witnessed its efficacy several times, and would have others test it in similar cases."

Dr. Peters says that Bromine causes inflammation of a transudative character in the larynx and trachea, with commencing formation of false membranes; violent inflammation of the fauces and œsophagus, and coating of them with *plastic lymph;* intense inflammation of the larynx and trachea, with *exudation of plastic lymph* in such abundance as quite to block up the air passages. He adds that it is rather more applicable to the inflammatory cases tending towards the larynx with sharp fever at the outset. Dr. J. P. Dake gives the following indications: "Soreness and smarting in the throat; ptyalism; hoarseness; rough, dry cough; sensation of contraction in the windpipe; fluent coryza; also nasal obstruction; epistaxis; earache; alternate chills and heat; violent inflammation of the mucous membrane of the fauces, œsophagus, also of the larynx and trachea; these parts are covered with a coagulable lymph which obstructs almost entirely the air passage. A dingy, brownish, granular, firmly-adhering exudation over the mucous membranes of the œsophagus." Dr. Trinks recommends it in severe inflammation of the fauces and œsophagus covering them with plastic lymph, also in severe inflammation of the larynx and trachea, with exudation of plastic lymph nearly closing these organs. Dr. Laurie recommends Bromine to be used in laryngeal diphtheria when Iodine has failed. Dr. McNeil advises Bromine "when the disease commences in the larynx and comes up into the fauces, and in some cases in which it runs down into the larynx and produces a croupy cough, with much rattling of mucus." Dr. W. C. Dake observes that "generally we have not had satisfactory results from its use," and Dr. Hughes says that "Bromine is the only rival of Kali bichromicum when diphtheria invades the larynx," yet his personal experience has not been favorable. Dr. Charles Neidhard, of Philadelphia, says, "I have been consulted in four or five cases of diphtheritic croup where Bromine was freely administered, in large and small doses, without any effect. They all died. In one or two of my own cases, it was also administered without benefit. It would seem that Bromine has not much effect in diphtheritic croup, nor in diphtheria generally." Personally, I have made little use of Bromine in this disease, partly because, as Dr. Bayes remarks, it is "an unmanageable medicine," and partly because I seldom saw grounds for its administration. I conclude that, though it is occasionally indicated in pseudo-membranous croup, it is not in homœo-

pathic *rapport* with diphtheria in any of its manifold phases, great care should be taken to preserve this remedy and the dilutions should be made each time it is used. "A gargle made with one drop of pure Bromine to six ounces of water has proved serviceable in diphtheria threatening to invade the larynx. It makes the false membrane brittle and brings it away, while it stimulates the subjacent mucous membrane."
—(*Bayes.*)

According to Dr. Peters, Ammonium causticum causes reddening of the nasal mucous membrane which is coated with an albuminous layer; reddening of the posterior surface of the epiglottis and of the entrance into the rima glottidis, which are covered with a false membrane; great redness of the whole trachea and bronchi, which are coated here and there with *membranous patches*. He adds that it may be used in diphtheria when the prostration and exhaustion are very great, and the disease tends to extend down into the larynx, trachea and air passages. Dr. J. P. Dake states that he has used this remedy with gratifying results, *by nasal inhalation*, but has not found benefit from the internal administration of the drug. Dr. McNeil observes that as this remedy has been used but little, we need further clinical provings (cures?) to clearly establish its province in diphtheria.

Dr. F. X. Spranger of Detroit reports the following cure in the *American Homœopathic Observer*, Vol. I : "Among the many cases that I have successfully treated with this medicine, I shall mention but one. It was a case of croupous diphtheria; a servant girl 20 years of age; corpulent, plethoric constitution. When first called to see her she had a croupous cough, which threatened suffocation every moment. On examination, found the lower part of the pharynx covered with a white pseudo-membrane extending down as far as could be seen. Patient was in the greatest agony, frequently jumping out of bed and gasping for breath. I dropped 15 drops of Ammonium causticum into a tumblerful of water, one-half-teaspoonful to be given every hour.

Left the patient soon afterwards, about 6 o'clock P. M. (The patient lived in the country.) While taking the first few doses she nearly strangled, deglutition being so difficult. Soon afterwards she began to get easier. Next morning I found the patient sitting up in bed, breathing freely. Had taken some broth; deglutition was very easy; the pseudo-membrane had entirely disappeared, and the patient was discharged cured the next day afterwards." Commenting on this, Dr. Oehme remarks that "though there is but one cure on record, yet we are forced to consider Ammon. caust. a great remedy in diphtheritis."

Excellent observers of the dominant school bear unwitting testimony to the homœopathicity of this remedy to the diphtheritic process. Thus Trendelenburg found that it was capable of causing the formation of false membranes in the trachea, and Dr. H. C. Wood confirms the observation. Delafond called "croup" into existence by means of Ammonia, and Oertel constantly insists on there being "no actual difference between croup as it ordinarily occurs, and that excited in the windpipe of a rabbit by means of Ammonia. The color and texture, the physical, chemical, and histological characteristics are identical."

Dr. Hughes advises Apis mellifica in the 'croupal' form of diphtheria "when a lower type of inflammation (as shown by a mere purple color of the parts) and much greater œdema are the first signs of the supervention of the croupous upon the catarrhal form of diphtheria, or of its primary onset." Dr. Oehme, whose work on the therapeutics of diphtheria is simply invaluable, remarks: "Because one physician has found Apis of no benefit in diphtheritis of the larynx, it does not follow that it will be thus in *all* cases, as we cannot expect *one* drug to be the *only* remedy for this disease." And he adds: "If we take into account the following symptoms: "Voice grew hoarse; breathing and swallowing very difficult; difficulty of swallowing not caused by the swelling of the throat, but by the irritation of the epiglottis; sensation as of a rapid swelling of the lining membrane of the air-passages ; rough voice ; speaking painful ; hoarse cough ; intense sensation of suffocation, could bear nothing about the throat ; hurried difficult respiration ; labored inspiration as in croup, etc.;" we see no reason why it should be neglected in these cases." In confirmation of these indications, I report the following case in the *American Observer*, Vol. XV: "On November 20th, 1877, I was called to W. S., a boy aged nine years. He had flying chills, followed by great heat with debility ; pulse 108 ; temperature in the axilla, $102\frac{1}{2}°$. The throat was very red with difficult deglutition and severe pains, felt even when not swallowing. I prescribed Apis mel., 5th dec. trit., one grain in eight teaspoonfuls of water, a teaspoonful every hour. At the same time I ordered the Grauvogel gargle, composed of equal quantities of spirits of wine and water, every two hours during the day, together with a diet exclusively of milk. For two days the situation remained almost unchanged, but on the morning of November 23d, a thick, yellowish, diphtheritic exudation covered the uvula, tonsils and pharynx, while the tongue had a thick, yellowish coating with inflamed papillæ and a high degree of fetor. The diphtheritic membrane was of the consistence of clotted cream, of a yellowish color, closely adherent to the subjacent mucous membrane, and of a fetid smell. The fever increased, and the morning temperature averaged $102\frac{1}{2}°$ and the evening $103\frac{1}{2}°$. No solid food could be taken, and small quantities of milk formed the sole nourishment. The Granvogel gargle was continued, though it caused intense pain each time it was used. The weakness and prostration increased to an alarming extent, and the characteristic bluish tint of the face was distinclty marked. The nostrils now became affected, and poured out a thin, fetid sanies. On November 28th the membrane extended lower down the pharynx, and on the following day the hoarse and croaking voice announced that the pharynx was at last involved, and this was confirmed by the stethoscope. Apis was now given in grain doses of the 5th dec. trit., dry on the tongue, every two hours. On November 30th the voice was entirely suppressed, with a hoarse and difficult cough, accompanied by the expectoration of small quantities of membrane. On December 1st the uvula began to shed its membrane, and during the five following days an astonishing amount of membrane was partly expectorated, partly vomited. Notwithstanding the very serious state of the larynx, no change was made in the remedy, except that the Grauvogel gargle was discontinued. On December 6th the tonsils and pharynx were almost clear of membrane, the voice returned, the laryngeal cough became softer, and the patient—very wan and prostrate—entered on convalescence. Throughout the entire progress of the disease, the patient presented Guernsey's key-note symptom, "*Puffiness about the eyes.*" Dismissed on December 8th, no remedy but Apis having been used. Since this case was reported, I have attended three others, strikingly similar to it, in which the same results were obtained from Apis.

Oehme gives the following excellent indications for Lachesis: "*The subjective symptom much severer than the objective;* violent pain in the throat; *extremely painful and difficult swallowing;* sensation of a foreign body in the throat, with stinging extending into the ear; urgency to swallow, and desire to hawk up something, with choking spells; dislike to have the throat touched; pale redness of the fauces; *exudate begins or is worse on the left side;* voice weak and hoarse; aphonia; cough causes pain; *fœtor oris;* fetid discharge from mouth and nose; *violent prostration,* even before the exudation; lassitude; weakness; pulse weak, small; perspiration cold, clammy; somnolency; delirium; *symptoms worse after sleep.*"

Dr. William Morgan says of this remedy: "This remarkable production of the animal kingdom may always be trusted as an auxiliary in removing that distressing and painful sensation of strangulation and suffocation, as if a cord were tied tightly round the throat, which marks certain forms of scarlet fever, the phlegmonous sore throat and diphtheria; indeed, I have never found Lachesis fail in this important symptom;" but as Dr. McNeil well remarks, "Lachesis is one of our most important remedies in both the laryngeal and septic forms," adding that "the indications are so clear that mistakes are inexcusable."

Dr. Ludlam, of Chicago, was the first to point out, on the authority of M. Laboulbéne, that the constitutional action of Tartar emetic will produce a pseudo-membrane upon the buccal, the laryngeal and the tracheal mucous surfaces. "The indications which, in my own experience, more frequently require this remedy in diphtheria, are sudden swelling of the cervical glands and tonsils, occurring in scrofulous children, who are predisposed to catarrhal or asthmatic affections; occlusion of the larynx or lower respiratory channels by excess of mucus, or of a feebly organized plasma, with cough, dysphagia, difficulty of breathing, gasping (which compels the patient to sit upright or to seek the open air); inclination to retching and vomiting, obstinate vomiting of a tenacious mucus without any considerable thirst; small circular patches, like smallpox pustules, in and upon the mouth and tongue; and also for evidences, of closure of the pulmonary air vesicles by solidification of effused serum (hepatization). It will sometimes serve a good purpose by promoting diaphoresis, and in exceptional cases will drive out the eruption, to the great relief of internal mucous surfaces. I reccommend you not to overlook the claims of this remedy in certain forms and varieties of the diphtheritic lesion. In particular, it seems applicable to many cases of diphtheria in which the abnormal throat and chest symptoms derive their chief characteristics from a prevalent influenza, or from an inherent predisposition on the part of the patient to catarrhal disorders of the respiratory mucous membrane."—(*Ludlam*). The writer has used Tartar emetic to a considerable extent, but thinks that it is more appropriate for pseudo-diphtheria than in the genuine disease. As to the dose, a grain of the 3d or 4th decimal trituration may be dissolved in half a cup of water and a teaspoonful given every one or two hours, or a small powder of the 5th or 6th trituration may be given dry on the tongue every two hours.

If the diphtheritic laryngitis should be a primary disease, good results may be expected from Aconite, provided it is given promptly and in material doses, certainly not higher than the 2d dec. dilution, but if it does not check the disease at once, some other remedy should be substituted for it.

Sir George Duncan Gibb recommends Sanguinaria "as an emetic in the croupal form of diphtheria." My own experience is that when the membrane is diffluent this remedy is effective, but not when the membrane is tough and closely adherent to the mucous membrane.

Dr. W. C. Dake, who has had excellent results in this disease, seems to consider that Spongia rivals Kali bichromicum. He advises "Spongia for paleness of the face and anxious features; stitches in the throat, great dryness of the larynx, with short, barking cough; difficult breathing, as from constriction of the larynx and trachea; pain in larynx when pressing upon it: hoarseness; dry cough, worse in the evening and toward morning from a tickling in the windpipe." Dr. Dake gives Spongia in the 1st decimal dilution, a dose every hour.

APHORISMS.

1. Diphtheritic croup is the development upon the larynx and trachea of the characteristic membrane of diphtheria, and, as a general rule, the disease originates by extension from the pharynx, it being a very rare thing to find it originate in the larynx.

2. In some epidemics, laryngeal diphtheria is so common as to give the characteristic features and name to the disease, while in other epidemics it is very rare.

3. It is more common in low, swampy lands, and on the margin of bodies of water, than on high and rolling land,

4. The proportion of cases of laryngeal diphtheria, as compared with the whole number of cases of diphtheria, varies from 15 to 67 per cent., and the mortality in cases of diphtheria in which the larynx is attacked, varies from 40 to 95 per cent.

5. Many of the croup epidemics of the eighteenth century, in England, Scotland and on the North American Continent, were epidemics of what would now be called diphtheritic croup.

6. When death takes place in diphtheria within a week, it is usually by extension of the disease to the larynx; when death takes place later, it is almost always the result of asthenia.

7. The characteristic membrane is of various textures, sometimes as soft as thick cream, sometimes like moist kid-leather.

8. The membrane may extend from the epiglottis to the minute ramifications of the bronchial tubes, and as it descends it becomes less and less consistent.

9. The prognosis is very unfavorable, even under enlightened homœopathic treatment, and the danger largely arises from the fact that the patient is suffering from a serious blood poisoning in addition to the local disease, and also that when the laryngeal complication appears the patient is already exhausted by the primary disease.

10. The younger the child the greater the danger of diphtheritic croup coming on in the course of diphtheria, and the younger the child the greater the danger when it does make its appearance.

11. As to therapeutics, Kali bichromicum heads the column, closely followed by Apis mellifica, Lachesis and Ammonium causticum. Of less importance, but still deserving of careful study, are Spongia, Iodine, Aconite, Sanguinaria, Bromine and Tartar emetic.

12. While tracheotomy is often the last reserve—and a

successful one too, *if not too long delayed*—in true croup, it is seldom admissible in diphtheritic croup.

13. The physician may sometimes be tempted to use an emetic for the purpose of removing the membrane from the larynx, but the relief is, at best, only temporary, and the irritation of the emetic action often intensifies the disease and hastens the fatal issue.

CHAPTER X.

SCARLATINAL CROUP.

Scarlatinal croup is a phase of disease to which exceedingly little reference is made in the medical writings of any school, and yet, though it is fortunately infrequent, it requires skill and promptitude more than any other complication of scarlatina. Objection may be made to any separate chapter on this subject, as the malady forms one phase of a general disease, and hence should be described with that disease. However, on account of the dangerous nature of the complaint, and also in view of the fact that no essay on the subject is contained in the literature of our school, I have thought it best to present the following chapter.

Scarlatinal croup, then, is a secondary inflammation of the larynx, occurring almost exclusively in the most malignant forms of scarlatina when the whole mass of fluids has been vitiated. It may originate by extension of the inflammatory irritation from the pharynx, though it sometimes appears when the pharynx is but little affected.

Scarlatinal croup is not a common phase of disease, for in the words of Professor Trousseau, "*scarlatina has no liking for the larynx."* It may appear in patients of any age, but it seems to me to be most frequent between the ages of four and eight. I have never noted it in infants, and all my patients, except two, were under ten years of age. Both sexes seem to be alike liable to the disease.

In many instances, scarlatinal croup originates by extension of the well-known sore throat of scarlatina, but in most of the cases I have observed, *exposure to cold* was the exciting cause. The illustrious Sydenham—doubtless encouraged by the success of his cool regimen in small-pox—thought that scarlatina patients ought to get up every day, even when the eruption was at its height. But scarlatina patients are much more susceptible to cold than small-pox patients; in fact, above all the eruptive fevers scarlatina needs to be guarded against cold. All my fatal cases originated in exposure to cold. One wilful nurse stripped a little patient to the skin at the height of scarlatina, and carried it about in a fireless kitchen for the purpose of "cooling the fever." Scarlatinal croup came on, and the case was hopeless when next seen. Another woman kept her little one, sick of scarlatina, in a well-warmed room during the day, but every night removed it to her own fireless bedroom situated at the extremity of a long, rambling farm-house, and this, too, during the month of February, 1868—the coldest part of the most severe Winter I ever saw in Canada. Here, too, the larynx was attacked with fatal result. In January, 1870, among other scarlatina patients I had one who made a fair recovery, though the type of disease was malignant. After I dismissed the case, the mother kept the cradle exposed to the cold air blowing in through an imperfectly-closed window, and fatal scarlatinal croup was the result.

Croup may come on during the early stages of scarlatina, or it may be one of sequelæ. It usually comes on insidiously and, amidst the anxiety of a serious disease, it may be unnoticed for a time. There is at first a very slight hoarseness, with muffled cough and a mingled gurgling and trilling sound in the larynx; after the cough the gurgling disappears for a time. These symptoms are frequently preceded by a slight chill, followed by heat of skin and accelerated pulse, but this may easily pass without remark. At first there is no dyspnœa, but soon marked difficulty of breathing comes on, and the dyspnœa indicates the degree of danger present, which is usually in precise proportion to this symptom. The patient involuntarily rises in bed and stretches out the head, while the eyes have an anxious and haggard expression. The cheeks are flushed and the eyes suffused. At this stage the tissues of the neck become swollen and infiltrated, and this, of course, increases the dyspnœa and hoarseness. There are no intermissions in this variety of croup; there is, however, a very slight remission in the morning, and usually a very severe exacerbation during the hours immediately before and after midnight. There is, in a majority of cases, a steady, onward march of the disease, the dyspnœa increases, the respiration becomes more stertorous, the cough, after becoming harsher, is finally suppressed, the strength fails, wild terrors and the ever-present feeling of suffocation prevent sleep, and finally the patient dies, comatose or convulsed. But, on the other hand, under the influence of a well chosen remedy, the dyspnœa may decrease, the cough may become less frequent and less hoarse, quantities of membrane may be vomited or swallowed, and the sleep of the patient then announces that the pressing danger has passed away. In another group of cases croup comes on suddenly and almost without warning. At one visit you leave your scarlatina patient doing well, and, when you next see him, the case is hopeless or almost hopeless.

The progress of this disease is very rapid, even more so than pseudo-membranous croup. Most of the fatal cases I have seen lived only from two to three days.

The false membrane of scarlatinal croup is thinner, softer and less adherent than the membrane of pseudo-membranous croup; at the same time, it is less uniformly spread over the affected part. It is grayish or of a yellow color, and is frequently associated with small quantities of pus, or it may be granular in texture and friable in consistence. But little fibrin enters into its composition and it rapidly decomposes. The subjacent mucous membrane is softened and of a dark purplish hue, while the sub-mucous areolar tissue is infiltrated; in fact, all the pathological appearances point to the localization of a degenerated blood disease. Professor Wood remarks that the membrane seldom extends, unless in small quantities, below the larynx.

In the great majority of cases the diagnosis is plain, for the history of the case must be investigated as well as the present state of the patient. The only cases in which there is reasonable ground for doubt are those which Trousseau denominates *defaced* scarlatina (scarlatine fraste), in which some of the most important symptoms of the malady are suppressed or non-existent. When, for example, there is no appearance of the characteristic eruption, but instead you have severe sore throat, with deposition of false membrane, it would be difficult to decide whether the disease was scarlatina or diphtheria, for a fetid smell exhales from the mouth and nostrils, the pulse is small and fluttering, the skin is pale and the temperature of the body is notably low. In such cases one of the best diagnostics would be the period at which albuminuria appeared, for, as is well known, in diphtheria it appears early in the disease, while in scarlatina it does not make its appearance till the case is far advanced. But in about one-fifth of the whole number of diphtheritic cases there is no albuminuria, and then the physician must look for other diagnostic points. There are two sources of fallacy in scarlatinal croup, to which I would direct special attention. The first of these will be found in the phenomena presented by a certain number of cases of scarlatina in which a quantity of matter in the posterior nares and upper part of the pharynx forms a mucous rhoncus which closely simulates croup. But here auscultation clears up the difficulty at once by showing that the larynx is not involved. In another set of cases the tumefaction of the neck is so great that it causes stertorous respiration, which bears a certain resemblance to croup. Here, too, auscultation is of some value, but a better diagnostic is the absence of the hoarse cough.

I look upon scarlatinal croup as being one of the most fatal of all the varieties of croup. It is more dangerous when it comes on at an advanced period of the course of scarlatina—say the tenth or twelfth day—than when it attacks at an early period. It is very dangerous when it arises by extension from the pharynx, but it is still more dangerous when it appears as an intercurrent inflammation, the result of exposure to cold. Tumefaction of the neck, if of great extent, is an unfavorable sign, and when coma or delirium appear there is little room for hope. Much, very much, depends upon prompt recognition of the disease and upon equally prompt therapeutics.

But one of the weak points about our knowledge of scarlatinal croup is that we have no well-defined treatment such as we possess in so many affections, and I regret that I can give but a few fragmentary hints derived entirely from personal experience. Here I cannot refrain from again pointing out the necessity of opposing the very beginnings of disease. "*Obsta principiis.*"

When recognized at an early period, Aconite is indicated in a majority of cases, but it should be given in the form tincture, as dilutions are merely a waste of invaluable time. I have great confidence in Sanguinaria and the confidence is derived from the fact that since I have used this remedy I have been much more successful than formerly.

A homely proverb says that "an ounce of prevention is worth a pound of cure," and I am strougly of the opinion that inunctions of olive oil are preventive of scarlatinal croup as well as of many of the complications and sequelæ of scarlatina. I use them in every case of scarlatina as follows: I direct one arm of the patient to be bathed lightly with tepid water, and then quickly dried, and, when thoroughly dry a small quantity of pure olive oil is rubbed over the limb. Then the other arm is treated in the same manner, and so on, till the entire person has been bathed and anointed. As a result the temperature is lowered, the irritation of the skin is allayed, and the liability to take cold is almost wholly removed.

APHORISMS.

1. Scarlatinal croup, fortunately not a common disease, appears in children of any age, and both sexes seem to be alike liable to it.

2. The disease may originate by extension from the pharynx, but it is most commonly caused by exposure to cold.

3. Scarlatinal croup is one of the most rapidly fatal of all the forms of croup, and it is more dangerous when it appears late in the course of scarlatina than when it comes on at an early period.

4. The leading remedies are Aconite, Sanguinaria and Kali bichromicum, and inunction with olive oil is the best prophylactic.

CHAPTER XI.

TRACHEITIS.

Is there such a disease as tracheitis? To read one series of medical authors, one would quite believe that there was such a disease, distinctly marked and well known, and as thoroughly understood as any malady in the nosological tables. To read another set of writers—quite as eminent and quite as well informed as the other—one would conclude from their brief remarks, and still more from the silence of some of them, that while all other parts of the human organism may be the victim of what the lamented Constantine Hering used to call "an *itis*," the trachea was the one happy spot which never knows what inflammation means. By one group of writers *tracheitis is considered to be synonomous with croup.* Sir Thomas Watson—the Macaulay of British Medicine—speaks of " another of Dr. Cullen's species of cynanche—viz.: *cynanche trachealis—tracheitis—croup;*" and Hasse—perhaps the most eminent of the Swiss pathologists—considers tracheitis and croup to be interchangeable terms. When such curious errors are made by the great lights of the school of medicine which has most zealously cultivated pathology and pathological anatomy, one does not wonder to see the author of the *Hydropathic Encyclopædia* follow in their wake.

George B. Wood, M. D., formerly Professor of Theory and Practice of Medicine in the University of Pennsylvania—certainly the greatest writer on disease that this continent has yet produced—offers the following remarks: "In a pathological account of the several portions of the air passages, it might be thought that the trachea would receive a separate consideration; but it is very seldom exclusively affected, offers no symptoms when inflamed which are not observed in other localities, and requires absolutely nothing peculiar in its treatment. The nomenclature which gives the title of tracheitis to croup is founded on a false assumption in relation to the special seat of that complaint. It is true that the trachea is generally affected in croup; but it is almost never exclusively affected; nor are the peculiar features of the disease essentially connected with that part of the respiratory passages. The symptoms and treatment of tracheitis are almost always merged in those of laryngitis and bronchitis." The fact that the trachea is very seldom exclusively affected, is not a good reason for passing over the disease altogether, as the same remark might be made respecting some other parts of the organism; and it is difficult to understand what is meant by the remark that it "offers no symptoms when inflamed which are not observed in other localities," seeing that tracheitis is quite different from the inflammations of kindred regions, and that the same remark might be made regarding almost any local inflammation—especially the inflammations of the encephalon. The remark that the disease "requires absolutely nothing peculiar in its treatment," is quite in place with therapeutics who are groping in the twilight of the so-called "physiological medicine," but it is repudiated by those who heal the sick in accordance with a great law of nature.

Tracheitis, then, may be defined to be an inflammation of the mucous membrane of the windpipe, usually arising from exposure to cold, and characterized by a croup-like cough with profuse secretion of mucus. This inflammation may be either primary or secondary; in the latter class it reaches the trachea by extension from the larynx. Although the inflammation may extend downwards from the larynx to the trachea, it is rare to find it extending upwards from the trachea to the larynx, and a similar remark may be made as to the relation between tracheitis and bronchitis. Indeed, it seems to be a general law that inflammations of the respiratory organs extend downwards.

Tracheitis is usually a catarrhal inflammation, though it is sometimes sthenic. Very seldom has the writer seen it of a pseudo-membranous nature, though it is well to bear in mind Rindfleisch's caution: "the development of a false membrane is connected in the closest manner with the catarrhal state, and constitutes the anatomical acme of the morbid process." It is, perhaps, never diphtheritic, though diphtheria may extend from the larynx to the trachea. This disease has no separate history, for, following Cullen's vicious nosology, it has almost invariably been confounded with croup, from which it differs in many important particulars. Almost all writers who have recognized the existence of tracheitis have remarked its infrequency, but a more careful examination, and especially the more frequent use of the stethoscope, would have proved that it is quite frequently met with. Boys are more subject to it than girls, and it is more frequent in fall and spring than at other seasons of the year.

Exposure to draughts of cold air is the most usual cause of tracheitis, but in children *wet feet* will usually be found to be the starting point of the disease. Children who have been confined to the house during the winter season are very apt to be attacked with laryngeal and tracheal inflammations when they first go out in spring, and nearly all the writer's cases occurred in March and April. Previous attacks form an undoubted predisposing cause of the disease, and the susceptibility is increased with each attack.

Tracheitis is usually preceded by premonitory symptoms resembling those of catarrh. The patient has more or less chilliness—not the distinctly marked chill of a sthenic inflammation, but a creeping, disagreeable feeling of chilliness, intermingled with heat—and this chilliness is followed in increased heat of the surface, with marked lassitude and loss of appetite. Felix von Niemeyer points out that this chilliness is rarely confined to a single rigor, thus forming an important point of distinction between the onset of a catarrhal and of an inflammatory fever. Shiverings recur with every little alteration of temperature, or on such slight exposure as changing the linen. A dull frontal headache is present, with throbbing of the temporal arteries, bruised pain in the limbs, and pain in the joints, increased by pressure or motion.

Sometimes there is only a slight irritation in the trachea, or a kind of tickling which provokes a short, hacking cough, but usually the cough is violent and paroxysmal. This cough is of three tolerably distinct varieties, according as the inflammation is confined to the windpipe, or, in addition, touches the larynx or the bronchi. When the inflammation is confined to the trachea, the cough is at first dry and spasmodic and of frequent recurrence, later there is an expectoration of thick, ropy mucus; when the larynx is implicated, the cough is hoarse, metallic and convulsive, while the breathing is loud and wheezing; when the disease invades the bronchi, the cough is dry at the commencement, but becomes looser in two or three days, with sputa of frothy mucus mixed with pus and sometimes streaked with blood. Prosser James remarks that "the voice will be unaffected so long as the larynx is not also involved;" but I have noted that even when the larynx is not affected the voice has a ringing and metallic sound which is quite distinct from the hoarse clangor of true croup. There is no real dyspnœa in tracheitis, owing to the large calibre of the affected organ as compared with that of the larynx or bronchi. The expectoration in simple tracheitis is more copious than in laryngitis, but much less abundant than when the bronchi are involved in the inflammatory action. Prosser James says that occasionally the expectoration appears in rings; I have never observed this phenomenon. As might be expected from the anatomical relations of the trachea and œsophagus, there is considerable dysphagia accompanied by a feeling of constriction; sometimes this is so considerable that the ingesta returns by the nostrils and mouth. The cause of this is obvious, for, as the lower extremity of the trachea is necessarily fixed, it follows that the upper extremity moves upwards with the larynx in swallowing, and, therefore, if the tracheal mucous membrane is inflamed, pain must be felt during deglutition. Morell Mackenzie denies the occurrence of dysphagia in the course of tracheitis, on the ground that though he has watched for the symptom he has never yet met with it, but a very slight acquaintance with the rules of evidence shows the futility of such an argument. In severe cases there is considerable swelling of the neck, but this is rare. The characteristic pain of tracheitis is a burning, stinging pain, aggravated by pressure and motion, and felt much lower down than the pain of laryngitis or croup.

Like croup, tracheitis usually developes itself during the night. The patient has usually been suffering from what was assumed to be a slight cold, and has retired to bed, apparently in good health, but in the night he awakes with the spasmodic cough and the loud and ringing voice, while at the same time the skin is hot and the face flushed. During the day the malady intermits, but on the approach of evening the paroxysm reappears in an aggravated form and with increased fever. The cough becomes more frequent and more annoying, and the thick, ropy mucus can be distinctly heard rattling in the windpipe. Should the disease be controlled by proper treatment, the cough diminishes and the characteristic burning and stinging pain disappears; the voice resumes its natural tones, and the fever subsides, leaving behind it languor and malaise. If the patient is old enough, free expectoration takes place, but in a majority of cases the secretions are brought up to the pharynx and then swallowed. Relapses are quite frequent, and the patient is ever after subject to the disease. In cases which terminate unfavorably, the bronchial tubes become involved, the cough increases in frequency, the expectoration becomes purulent and profuse, the appetite fails, and the body emaciates, till the child sinks into its grave with a group of symptoms closely resembling those of phthisis pulmonalis.

Tracheitis of a sthenic nature lasts from five to seven days, and is more easily cured and is less likely to become chronic than if it were catarrhal in its nature. It is also less likely to invade the bronchi. Pseudo-membranous and diphtheritic tracheitis are rarely found, save as parts of one general disease from which the patient has but little chance of recovery. Catarrhal tracheitis is the most common variety, and so far as my experience goes, is comparatively easy of cure, but it often recurs and is apt to become chronic. In a great majority of cases it extends to the bronchial tubes, and this bronchitis is apt to become chronic. Chronic tracheitis may last for many months, and, as in one case which I observed, may linger for a year, getting alternately better and worse, till at last recovery takes place. Such cases may be complicated with chronic bronchitis, when, as has been remarked, the case will strongly resemble consumption.

Usually the thermometer shows an elevation of temperature, say from one to two degrees above health, at the commencement of the illness, while further on the temperature may be normal or nearly so. I have attended a number of cases in which the temperature remained normal throughout, and again, as von Niemeyer points out, during the intervals of the shiverings the patient may experience a sensation of burning heat, without any indication from the thermometer of an actual increase of temperature.

The laryngoscope can very rarely be used with young children, but the stethoscope gives equally valuable results. At the commencement of the disease sibilant râles are heard all over the tracheal region, with increased vocal resonance. When secretion takes place, large mucous râles replace the sibilant ones, and when the tracheal mucous membrane is swollen at some particular point a well-marked, sonorous rhoncus is heard.

Post mortem examinations, as well as examinations with the laryngoscope during life, show that in this disease the lining membrane of the trachea is of an intense scarlet redness, deepening into purple in very severe cases. The entire membrane is made up of a congeries of minute florid blood-vessels, so closely packed together as to give the characteristic scarlet hue. At the same time, the follicles are enlarged and very prominent, appearing like minute red points projecting from the inflamed surface, which has a kind of glazed appearance. These follicles form the source of the thick and ropy mucus which is coughed up. Ulcers are not common, and softening is rarely or never present, either in this disease or in laryngitis. Prosser James points out that when infiltration takes place it is more apt to be fibrinous than serous. In chronic tracheitis the redness is less than in the acute form, and the vessels are often varicose; the mucous membrane is often hypertrophied, indurated and ulcerated.

The diagnosis is comparatively easy. There is, of course, nothing characteristic in the fever which precedes the attack, for that is merely the fever common to all local inflammations. More characteristic are the burning, stinging pain, with its peculiar anatomical seat, the clear and ringing cough, with the gurgling of mucus in the windpipe, and in severe cases, the dysphagia, with constriction and swelling.

The prognosis in catarrhal or sthenic tracheitis is almost uniformly favorable—differing widely in this respect from laryngitis or brochitis. The principal danger is the disposition to become chronic and to extend to the bronchial tubes. Still, one can easily conceive that there would be danger if a young infant of feeble vitality should be attacked with catarrhal tracheitis with a very copious secretion of mucus.

The room in which the patient sleeps should be nearly of the same temperature as that which he occupies during the day, and while draughts should be carefully avoided, ventilation should be as carefully maintained. Warm moisture should be added to the atmosphere of the room in severe cases, and great care should be taken when the patient first goes into the open air after recovery.

Aconite must ever be the first remedy thought of if the physician is called at or near the commencement of the illness. It is adapted to the etiology of the disease, for exposure to wet and cold is by far the most influential factor in its causation. If, however, the physician is not called during the first forty-eight hours, Aconite is of little or no avail and another remedy must be selected according to the symptoms present. I have generally used Aconite from the mother-tincture to the 3d decimal trituration, and see no necessity for going higher in the scale.

In the catarrhal form no remedy equals Sanguinaria, which, moreover, is able to prevent extension to the bronchial tubes, and also the recurrence of the disease. I am thoroughly convinced that when patients subject to croup or tracheitis are treated with Sanguinaria, they lose the predisposition to these diseases. Usually after a few preliminary doses of the inevitable Aconite, I give Sanguinaria with unvarying success, though exceptional cases occur which demand other remedies.

Tartar emetic is indicated when the cough is very frequent with audible rattling of mucus in the windpipe and bronchi —the mucus is tough, white and copious. The larynx is little affected, but inclination to vomit is often present. Threatening paralysis of the lungs is best met by this remedy. Tartar emetic acts best in repeated doses of the 3d or 4th decimal trituration.

Mercurius solubulis is an excellent remedy in catarrhal tracheitis; dry, distressing cough, usually recurring at night and racking the entire frame. The patient is sensitive to cold, chills alternate with paroxyms of burning heat, and the tracheitis is merely the most prominent part of a general catarrhal fever. I have generally given this remedy in powders of the 6th decimal trituration, dry on the tongue.

Spongia is indicated when the cough is hoarse, ringing and hollow, with labored and wheezing breathing. The cough is distinctly paroxysmal, and is usually without expectoration. Spongia acts in all dilutions; I usually give the 6th decimal trituration.

Hepar is very like Spongia, differing from it in the large quantity of mucus present. The cough is hoarse and barking, and a suffocative feeling is almost constantly present. Hepar acts well in all dilutions; I prefer the 12th decimal trituration.

Sulphur is frequently indicated towards the close of the disease when a dry cough remains with feeling of constriction in the chest, worse after eating or during deep inspiration. The 30th is the most suitable preparation.

Aphorisms.

1. Tracheitis exists as a separate disease, though, as it often co-exists with laryngitis and bronchitis, its existence has been questioned.

2. The leading symptoms are a burning, stinging pain in the windpipe, with dysphagia and local swelling. These symptoms are preceded by fever and accompanied by dry, spasmodic cough and ringing voice.

3. The prognosis is almost uniformly favorable, but the disposition to become chronic and to extend to the bronchi should be carefully guarded against.

4. The homœopathic remedies are Aconite, Sanguinaria, Tartar emetic, Mercurius solubulis, Spongia, Hepar and Sulphur. Sanguinaria removes the predisposition to the disease.

INDEX.

	PAGE
Aconite, in acute coryza,	29, 37
in purulent coryza,	49
in spasm of the glottis,	89
in acute catarrhal laryngitis,	127
in acute œdametous laryngitis,	147
in spasmodic croup,	164
in pseudo-membranous croup,	218
in diphtheritic croup,	286
Ailanthus, in chronic coryza,	63
Allium cepa, in acute coryza,	34
Alumina, in chronic coryza,	58
in scarlatinal croup,	292
in tracheitis,	301
Ammonium carbonicum, in acute coryza,	36
Ammonium causticum, in diphtheritic croup,	281
Antimonium crudum, in chronic coryza,	62
Apis mellifica, in purulent coryza,	49
in acute œdematous laryngitis,	143
in diphtheritic croup,	282
Argentum nitricum, in purulent coryza,	47
in chronic coyrza,	57
in acute catarrhal laryngitis,	131
Asafœtida, in chronic coryza,	63
Arsenicum album, in acute coryza,	32
in spasm of the glottis,	91
in acute catarrhal laryngitis,	130
in acute œdematous laryngitis,	148
Arsenicum iodatum, in chronic coryza,	62
Arum triphyllum, in acute coryza,	36
Aurum metallicum, in chronic coryza,	57
Baryta carbonica, in chronic coryza,	59
Belladonna, in acute coryza,	33
in purulent coryza,	49
in spasm of the glottis,	92
in acute catarrhal laryngitis,	129
Berberis, in chronic coryza,	63
Bouchut, on the use of canulæ in acute coryza,	37
Bromine, in spasm of the glottis,	108
in pseudo-membranous croup,	225
in diphtheritic croup,	279

INDEX.

Bryonia, in acute coryza,	36
in acute catarrhal laryngitis,	130
in pseudo-membranous croup,	232
Calcarea carbonica, in chronic coryza,	55
Camphor, in acute coryza,	30
Carbo vegetabilis, in acute coryza,	36
Chamomilla, in acute coryza,	33
in spasm of the glottis,	109
Chlorine, in spasm of the glottis,	99
Coryza, acute,	25
etiology of,	26
nature of,	27
symptomatology of,	27
progress of,	29
therapeutics of,	29
general management of,	37
aphorisms of,	38
Coryza, chronic.	51
varieties of,	51
etiology of,	51
symptomatology of,	52
thermometry of,	53
diagnosis of,	53
prognosis of,	54
general management of,	54
therapeutics of,	54
aphorisms of,	63
Coryza, purulent,	39
etiology of,	41
symptomatology of,	42
thermometry of,	44
pathological anatomy of,	44
prognosis of,	46
general management of,	46
therapeutics of,	47
aphorisms of,	50
Corallia rubra, in spasm of the glott's,	108
Croup, diphtheritic,	235
history of,	236
etiology of,	241
symptomatology of,	244
pathological anatomy of,	249
prognosis of,	266
tracheotomy in,	268
general treatment of,	274
therapeutics of,	275
aphorisms of,	286

INDEX. 305

Croup, pseudo-membranous,	171
definition of,	173
etiology of,	177
symptomatology of,	190
progress of,	199
thermometry of,	201
physical diagnosis of,	202
essential nature of,	204
pathological anatomy of,	206
diagnosis of,	210
prognosis of,	214
tracheotomy in,	216
therapeutics of,	218
aphorisms of,	233
Croup, scarlatinal,	288
definition of,	288
etiology of,	289
symptomatology of,	289
pathological anatomy of,	290
diagnosis of,	291
prognosis of,	292
therapeutics of,	292
aphorisms of,	293
Croup, spasmodic,	154
etiology of,	157
physical diagnosis of,	162
pathological anatomy of,	162
diagnosis of,	163
prognosis of,	163
therapeutics of,	163
general treatment of,	168
aphorisms of,	170
Cyclamen, in acute coryza,	36
in chronic coryza,	62
Cuprum metallicum, in spasm of the glottis,	106
Dulcamara, in acute coryza,	36
Dunham, Dr. Carroll, on chlorine in spasm of the glottis,	100
Euphrasia, in acute coryza,	34
Gelsemium, in spasm of the glottis,	95
Graphites, in chronic coryza,	60
Hayward, on the local use of Aconite in acute coryza,	37
Hepar sulphuris, in acute coryza,	32
in chronic coryza,	62
in acute catarrhal laryngitis,	128
in spasmodic croup,	166
in tracheitis,	302

Hering, Constantine, on the etiology of acute coryza,	38
Hippocrates, on the general treatment of acute coryza,	37
Hydrastic canadensis, in chronic coryza,	62
Hyosciamus, in acute catarrhal laryngitis,	131
Ipecacuanha, in acute coryza,	36
in spasm of the glottis,	94
in acute catarrhal laryngitis,	131
Ignatia, in spasm of the glottis,	106
Iodine, in chronic coryza,	62
in spasm of the glottis,	98
in acute œdematous laryngitis,	148
in pseudo-membranous croup,	221
in diphtheritic croup,	278
Kali bichromicum, in chronic coryza,	56
in pseudo-membranous croup,	227
in diphtheritic croup,	275
Kali carbonicum, in chronic coryza,	60
Kali hydrodicum, in chronic coryza,	61
Lachesis, in chronic coryza,	60
in spasm of the glottis,	107
in acute catarrhal laryngitis,	130
in acute œdematous laryngitis,	148
in diphtheritic croup,	284
Laryngitis, acute catarrhal,	116
nature of,	117
etiology of,	118
symptomatology of,	118
thermometry of,	121
pathological anatomy of,	122
aphorisms of,	123
prognosis of,	124
general treatment of,	124
therapeutics of,	125
aphorisms of,	127
Laryngitis, acute œdematous,	132
nature of,	132
varieties of,	133
symptomatology of,	136
progress of,	137
pathological anatomy of,	139
diagnosis of,	140
prognosis of,	142
therapeutics of,	143
operative interference in,	149
aphorisms of,	152
Laurocerasus, in spasm of the glottis,	109

INDEX. 307

Lobelia, in spasmodic croup,	168
Lycopodium, in chronic coryza,	59
Meigs, Dr. Chas. D., on the flannel cap in acute coryza,	37
Mercurius iodatus, in chronic coryza,	62
Mercurius solubilis, in acute coryza,	31
in acute catarrhal laryngitis,	130
in tracheitis,	301
Moschus, in spasm of the glottis,	88
Nitric acid, in purulent coryza,	48
in chronic coryza,	61
Nux Vomica, in acute coryza,	30
in spasm of the glottis,	108
in acute catarrhal laryngitis,	131
Opium, in spasm of the glottis,	108
Phosphorus, in acute catarrhal laryngitis,	130
in spasmodic croup,	167
in pseudo-membranous croup,	232
Plumbum, in spasm of the glottis,	107
Pulsatilla, in acute coryza,	35
in spasm of the glottis,	108
in acute catarrhal laryngitis,	131
Sanguinaria canadensis, in acute coryza,	36
in spasm of the glottis,	90
in acute catarrhal laryngitis,	127
in acute œdematous laryngitis	145
in pseudo-membranous croup,	228
in diphtheritic croup,	286
in scarlatinal croup,	292
in tracheitis,	301
Sambucus, in acute coryza,	36
in spasm of the glottis,	87
in acute catarrhal laryngitis,	131
Scald throat,	138
Sepia, in chronic coryza,	58
Silicia, in chronic coryza,	55
Spasm of the glottis,	64
nature of,	65
etiology of,	68
symptomatology of,	77
mode of death in,	87
therapeutics of,	87
general treatment of	110
chloroform in,	111
tracheotomy in,	111
lancing of the gums in,	112
diet in,	112
aphorisms of,	114

Spongia, in spasm of the glottis, . 109
    in acute catarrhal laryngitis, 128
    in acute œdematous laryngitis, 148
    in spasmodic croup, . 165
    in diphtheritic croup, 286
    in tracheitis, ... 301
Stannum metallicum, in chronic coryza, 62
Sulphur, in chronic coryza, . . 54
    in spasm of the glottis, 109
    in tracheitis, . . . 302
Tartar emetic, in acute catarrhal laryngitis, 129
        in pseudo-membranous croup, . 231
        in diphtheritic croup, . . 285
        in tracheitis, . 301
Tracheitis, ... 294
    definition of, . 295
    etiology of, . . . 296
    symptomatology of, . . 296
    thermometry of, . 299
    pathological anatomy of, . 299
    diagnosis of, . 299
    prognosis of, . . . 300
    general treatment of, . . 300
    therapeutics of, . . . 301
    aphorisms of, ... 302
Veratrum album, in spasm of the g ottis, . . 109
Williams, Dr. C. J. B., on the thirst cure in acute coryza, 37
Zincum metallicum, in spasm of the glottis, . . 108

www.ingramcontent.com/pod-product-compliance
Lightning Source LLC
Chambersburg PA
CBHW031329230426

43670CB00006B/285